Bricklaying

LEVEL 1 DIPLOMA (6705)

▌ Mike Jones

The City & Guilds textbook

HODDER
EDUCATION
AN HACHETTE UK COMPANY

Orders: please contact Hachette UK Distribution, Hely Hutchinson Centre, Milton Road, Didcot, Oxfordshire, OX11 7HH. Telephone: +44 (0)1235 827827. Email education@hachette.co.uk Lines are open from 9 a.m. to 5 p.m., Monday to Friday. You can also order through our website: www.hoddereducation.co.uk

ISBN: 978 1 3983 1936 3

© The City and Guilds of London Institute and Hodder & Stoughton Limited 2021

First published in 2021 by
Hodder Education,
An Hachette UK Company
Carmelite House
50 Victoria Embankment
London EC4Y 0DZ

www.hoddereducation.co.uk

Impression number   10 9 8 7 6 5 4 3 2

Year   2025 2024 2023

Cover photo © PHILL THORNTON PHOTO – stock.adobe.com

Illustrations by Integra Software Services Ltd, Mike Jones and Andy Buckle

Typeset in India by Integra Software Services Ltd

Printed and bound by CPI Group (UK) Ltd, Croydon, CR0 4YY

A catalogue record for this title is available from the British Library.

Contents

About your qualification

This textbook supports learners working towards the City & Guilds Level 1 Diploma in Bricklaying (6705-13).

A Level 1 qualification can give you a solid foundation on which to build your career. The qualification introduced in this textbook provides the knowledge and skills needed to build your confidence and ability on your journey towards qualification as a bricklayer.

Key learning areas are set out in the qualification as units of study which cover the basic requirements for a bricklayer's work. Subject matter includes:

- the types of construction documents used in the workplace
- how to interpret and apply information presented on working drawings
- understanding construction principles and the different parts of buildings
- the methods used to set out basic structures and ensure they are at the correct height levels
- how to build solid walls and cavity walls using bricks and blocks.

An essential chapter focuses on health and safety on construction sites and gives you vital insights into how you can protect your own welfare and that of others.

How your knowledge and skills will be assessed

Examinations for each unit of the qualification take the form of multiple choice question papers set by City & Guilds. All the subject points covered in the qualification are covered in this book, and at the end of each chapter a series of practice questions are provided to test your knowledge and give you an idea of what to expect in your exams.

Many of the unit exams include a practical assignment, and throughout this book there are industry tips and step by step features that get you familiar with the skills, practical abilities and industry awareness you will need to tackle the assignments with confidence.

Continuing your learning journey

On attaining a Level 1 qualification, the natural progression is to study at Level 2. This level of qualification is widely accepted as the requirement for working efficiently and productively on site. There are many references in the chapters of this textbook that give you an idea of the additional subject matter that you can learn at Level 2.

In the longer term, you could choose to progress to Level 3 to study more complex technical construction subjects with the eventual possibility of moving to a supervisory position on site.

Acknowledgements

We would like to thank everyone who has contributed to City & Guilds photoshoots. In particular, thanks to: Andrew Buckle (photographer), Paul Reed, Akeem Callum, Frankie Slattery, Wahidur Rahman and all of the staff at Hackney Community College.

Contains public sector information published by the Health and Safety Executive and licensed under the Open Government Licence.

From the author

Although the author of a textbook writes the words and thinks about how best to explain things, there are many skilled and hardworking people who make a book like this possible.

My thanks to the copy-editors, proofreaders, illustrators and editors who have put so much time and effort into making this new textbook a reality.

I'm especially grateful to Tom Stottor who gave his valuable time discussing and exploring ideas on how to characterise this new Level 1 textbook, and then got the production process in motion with such enthusiasm.

Many thanks to Matthew Sullivan who has been so supportive and provided lots of suggestions and ideas on how to shape each chapter of the book into something of real value for Level 1 learners.

Working with Imogen Miles once more has again demonstrated her skills in laying out the book to be visually appealing as well as of practical value in her work as Senior Desk Editor. Thank you.

Finally, thanks to my long-suffering wife Sue who has patiently supported me and made sure I get enough cups of tea and slices of home-made cake to keep me going!

Mike Jones, 2021

About the author

Bricklaying and the construction industry have been central to my working life over a long period. I have worked as a skilled tradesman, a supervisor and a site manager on projects ranging from small extensions to multi-storey contracts worth millions of pounds. For a number of years, I employed a small team of skilled workers in my own construction company, working on contracts for selected customers.

The skills I developed over the years allowed me to design and successfully build my own family home in rural Wales, which I have always viewed as a highlight of my construction career. Relatively few people are able to have that privilege.

After over 30 years working on construction sites, I was fortunate to move into the education sector, first as an NVQ trainer and assessor for three years and subsequently as a college lecturer. I worked for ten years in the Brickwork section at Cardiff and Vale College in South Wales where I taught bricklaying and other construction skills and became section leader.

Since leaving my college post, I have maintained my links with training and education, producing teaching and learning resources for City & Guilds along with my work as a technical author writing bricklaying textbooks. I am pleased to be currently involved in the writing, reviewing and editing of resources for bricklaying qualification examinations.

Working 'hands-on' in the construction industry followed by my work in the education sector has been very rewarding. My aim during my time teaching others has been to impart to learners the great job satisfaction that can be gained from becoming a skilled practitioner. Put maximum effort into developing your skills and knowledge, and you will be able to take full advantage of career opportunities that come your way.

Mike Jones, 2021

Picture credits

The Publishers would like to thank the following for permission to reproduce copyright material.

Fig.1.1 © Ionia/stock.adobe.com; Fig.1.9 © Bannafarsai/stock.adobe.com; Fig.1.10 © Northcot Brick Ltd; Fig.1.14 © Smileus/Shutterstock.com; Fig.1.15 © Jesus Keller/Shutterstock.com; Fig.1.17 © Tomislav Pinter/Shutterstock.com; Fig.1.19 © El Roi/123 RF.com; Fig.1.20 © Alena Brozova/Shutterstock.com; Fig.1.21 © Maksim Safaniuk/123 RF.com; Fig.1.22 © Avalon/Construction Photography/Alamy Stock Photo; Fig.1.24 © Annelie Krause/123RF; Table 1.1 photo © Liliya Trott/stock.adobe.com; Fig.1.26 © Elroi/stock.adobe.com; Fig.1.27 © Kadmy/stock.adobe.com; Fig.1.32 © Ungvar/stock.adobe.com; Fig.1.36 © Kruraphoto/stock.adobe.com; Fig.1.37 © Stieberszabolcs/123RF; Fig.1.46 © SteF/stock.adobe.com; Fig.1.47 © Kasipat/stock.adobe.com; Fig.1.48 © NorGal/stock.adobe.com; Fig.1.50 © Michaeljung/stock.adobe.com; Fig.1.51 © Gorodenkoff/stock.adobe.com; Fig.1.52 © Rido/stock.adobe.com; Table 1.2 (top to bottom): © Rawpixel.com/stock.adobe.com, © Goodluz/stock.adobe.com, © Budimir Jevtic/stock.adobe.com, © Eric/stock.adobe.com; Fig.1.53 © Monkey Business/stock.adobe.com; Fig.2.1 © Birgit Reitz-Hofmann/stock.adobe.com; Fig.2.2 © Bijoy/stock.adobe.com; Fig.2.3 © SeagullNady/stock.adobe.com; Fig.2.4 © Luca pb/stock.adobe.com; Table 2.1 (top to bottom): © Stanley Black & Decker, City & Guilds, courtesy of Axminster Tool Centre Ltd., © Faithfull Tools, © Faithfull Tools, picture courtesy of Chestnut Products; Fig.2.12 City & Guilds; Fig.2.13 © Faithfull Tools; Fig.2.15 © York Survey Supply Centre Ltd; Fig.2.22 © STABILA; Fig.2.23 © 2020 Screwfix Direct Limited; Fig.2.24 © Lightfield Studios/stock.adobe.com; Fig.2.25 © 123RF; Fig.2.26 © 123 RF; Fig.2.27 © Roman_23203/stock.adobe.com; Fig.3.1 © Kings Access/stock.adobe.com; Fig.3.2 © Jahanzaib Naiyyer/Shutterstock.com; Fig.3.4 © Monkey Business/stock.adobe.com; Fig.3.5 © Elenathewise/stock.adobe.com; Fig.3.6 © Cavan/stock.adobe.com; Fig.3.8 © Bokic Bojan/Shutterstock.com; Fig.3.10 © Toeyfatboy/123 RF.com; Fig.3.11 © Dainis/Shutterstock.com; Fig.3.12 © Stephen Rees/Shutterstock.com; Fig.3.14 © Everbuild - A Sika Company; Fig.3.15 © Bostik Ltd; Fig.3.16 © Hanson UK; Table 3.2 (top to bottom): photo courtesy of MARSHALLTOWN, photo courtesy of MARSHALLTOWN, photo courtesy of MARSHALLTOWN, courtesy of Axminster Tool Centre Ltd., © STABILA, © Faithfull Tools, © Stanley Black & Decker, © Faithfull Tools, © Olympia Tools (UK) Limited, © Faithfull Tools; Table 3.3 (top to bottom): © Lost Mountain Studio/Shutterstock.com, © Eduardo/stock.adobe.com, © objectsforall/Shutterstock.com, © Faithfull Tools, © 2020 Screwfix Direct Limited; Fig.3.18 © Andrey Kuzmin/stock.adobe.com; Fig.3.19 © Xavier de Canto/Construction Photography/Avalon/Hulton Archive/Getty Images; Fig.3.20 © viphotos/Shutterstock.com; Fig.3.21 © Pavilion Construction Ltd; Fig.3.22 © Nattasak/stock.adobe.com; Fig.3.27 © UVEX SAFETY (UK) LTD; Fig.3.28 © Solidmaks/stock.adobe.com; Fig.3.29 © George Dolgikh/stock.adobe.com; Fig.3.30 © Virynja/stock.adobe.com; Fig.3.31 © Dmitry Kalinovsky/Shutterstock.com; Fig.3.32 © Saint-Gobain; Fig.3.33 © Stefan1179/stock.adobe.com; Fig.4.1 © Jahanzaib Naiyyer/Shutterstock.com; Fig.4.2 © Andrey Kuzmin/stock.adobe.com; Fig.4.3 © daseaford/Shutterstock.com; Fig.4.4 City & Guilds; Fig.4.5 © Pormezz/stock.adobe.com; Fig.4.6 © Cineberg/stock.adobe.com; Fig.4.7 © Northcot Brick Ltd; Fig.4.8 © Pixelrain/stock.adobe.com; Fig.4.9 © Bokic Bojan/Shutterstock.com; Fig.4.11 © Julija Sapic/Shutterstock.com; Fig.4.12 © Rob hyrons/stock.adobe.com; Fig.4.14 City & Guilds; Fig.4.15 © Avalon/Construction Photography/Alamy Stock Photo; Fig.4.16 © Everbuild - A Sika Company; Table 4.2 (top to bottom): photo

How to use this book

Throughout the book you will see the following features.

Industry tips are particularly useful pieces of advice that can assist you in your workplace or help you remember something important.

INDUSTRY TIP

You may hear the concrete slab used for a solid ground floor referred to as the 'oversite' concrete.

Key terms, marked in bold purple in the text, are explained to aid your understanding. (They are also explained in the Glossary at the back of the book.)

KEY TERM

Joists: parallel timber beams spanning the walls of a structure to support a floor or ceiling

Health and safety boxes provide important points to ensure safety in the workplace.

HEALTH AND SAFETY

Never use powered cutting equipment unless you have been properly trained and certified as competent.

Activities help to test your understanding.

ACTIVITY

Search online for 'surveyor's tape measures' and make a list of all the different lengths of measuring tape you can find.

Improve your maths items provide opportunities to practise or improve your maths skills.

Improve your English items provide opportunities to practise or improve your English skills.

At the end of each chapter there are some **Test your knowledge** questions. These are designed to identify any areas where you might need further training or revision. Answers are provided at the back of the book.

CHAPTER 1

PRINCIPLES OF CONSTRUCTION, INFORMATION AND COMMUNICATION

INTRODUCTION

This chapter considers the tried-and-tested ideas, methods and rules that are used to successfully design and build a structure. These are called 'principles', and they are applied to make sure that a building design is safe, efficient and comfortable for the users and does not damage the environment over time.

Building structures that are safe and long lasting is a complex process, involving teams of personnel who need to share information and communicate effectively.

By developing a good understanding of construction principles and learning to interpret and communicate information well, you can contribute more productively to an industry that has a significant impact on everyday life.

LEARNING OUTCOMES

After reading this chapter, you should:
1 know how to identify information used in the workplace
2 know how construction can affect the environment
3 know about the construction of different parts of buildings
4 know how to calculate quantities of materials
5 know how to communicate effectively in the workplace.

1 IDENTIFYING INFORMATION USED IN THE WORKPLACE

To contribute to the construction process, a bricklayer must be able to identify correctly and interpret accurately information provided by a range of drawings and written documents. Even a small project may depend on many different sources of information to be completed successfully.

In this chapter, we will consider the drawings and documents that are used to understand the different parts (or **elements**) of a building. These documents may contain sensitive information, for example, details about a government project or a bank building, and should therefore be stored carefully and kept secure at all times. Even if documentation does not relate to a sensitive project, it should be looked after carefully, since a project cannot be completed without it.

KEY TERM

Elements (of a building): the main parts of a building – the foundation, floors, walls and roof

1.1 Drawings

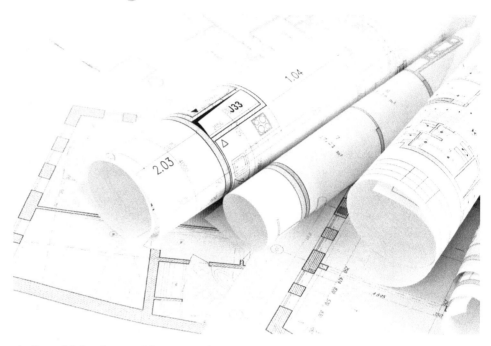

▲ Figure 1.1 Drawings used for construction

Drawings are used in construction to communicate a great deal of information, for example:

- the position and size of a building and its foundation
- internal features
- important details about specific parts of the structure
- the position of doors and windows
- the appearance of the building from the front, back and sides
- the shape of the roof.

Drawings can be referred to as contract documents, which means that they are legally binding. If a bricklayer decides to change details on a drawing without being authorised to do so, there could be serious legal consequences.

KEY TERM

Demolish: to deliberately destroy something (such as a building)

INDUSTRY TIP

There have been cases where the position of a building on a plot has been moved during construction, for seemingly good reasons. However, if these changes mean that the building no longer conforms to the approved legally binding drawing, the courts may give an order for the building to be **demolished**.

Block plan

When a building is designed, thought must be given to how its proposed position will affect existing buildings in the locality. For example, it could block out natural light falling on nearby properties or spoil the appearance of the neighbourhood.

Factors like these must be considered carefully, and a block plan is produced during planning to show the proposed development in relation to surrounding properties.

▲ Figure 1.2 Block plan

ACTIVITY

Look carefully at the arrangement and layout of buildings in the area where you live. Are any buildings shaded by others? Do you think they are too close together or are they spaced in an attractive manner?

Make a rough **plan view** sketch of the nearest six buildings to where you live (use the block plan in Figure 1.2 to help you) and suggest improvements to the existing layout.

KEY TERMS

Plan view: a view from directly above the subject – a 'bird's-eye' view

Scale: when accurate sizes of an object are reduced or enlarged by a stated amount

Ratio: the amount or proportion of one thing compared to another

When producing drawings, an architectural technician or a draughtsperson will draw a building to **scale**. This means that a large structure can be represented with accurate proportions on a document that is much smaller, which is more manageable than a full size drawing. Scale is shown by a **ratio** such as 1 to 10. This is usually written in the form '1:10'.

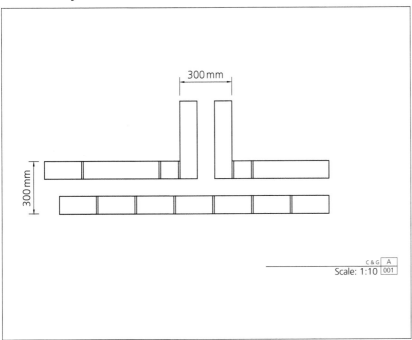

300 mm

300 mm

PROJECT
<C & G >
PROJECT NO.
<001>
ISSUE
<24/08/2020>
DRAWN BY
MJ

Plan view junction cavity wall

C & G   A
Scale: 1:10   001

▲ Figure 1.3 A drawing showing a scale of 1:10

In the case of a block plan, the ratio will usually be 1:1250 or 1:2500. If a block plan is drawn to a scale of 1:1250, a bricklayer would need to place 1250 drawings side by side to equal the size of the real thing.

A block plan is an essential overview of the proposed project, allowing the viewer to interpret important information about the location and positioning of the building. It usually shows individual plots and road layouts on the site as simple outlines without detailed dimensions.

Site plan

A site plan provides more detail than a block plan and shows the proposed building location in relation to the property boundary, so that the building can be accurately set out and positioned on a plot or site. The usual scale for a site plan is 1:200 or 1:500, which allows for this extra detail to be shown.

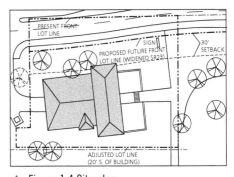

PRESENT FRONT
LOT LINE

SIGN
PROPOSED FUTURE FRONT
LOT LINE (WIDENED SR23)

30'
SETBACK

ADJUSTED LOT LINE
(20' S. OF BUILDING)

▲ Figure 1.4 Site plan

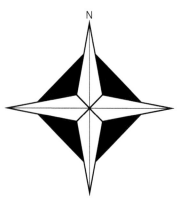

▲ Figure 1.5 Northing symbol

INDUSTRY TIP

You may come across alternative symbols to show north on different drawings, such as the example shown in Figure 1.8.

Site plans often show:

- dimensions for access roads and drives on the plot or site, so that their position in relation to the new building and existing highways or footpaths can be established accurately
- drainage pipe runs for the new development, along with other services such as water, electricity, gas and telecoms
- the proposed position of new trees and shrubs, which are sometimes required as a condition of the planning permission for a new development.

A site plan will usually show which direction is north (often referred to as 'northing' or 'north point'), so that the building can be orientated correctly.

ACTIVITY

Try to establish the direction of north from the front door of where you live. (Hint: in the UK, the sun will be due south at midday.)

Using Figure 1.4 as a guide, produce a plan view sketch of the building you live in. Show north using the symbol shown in Figure 1.5.

Section drawings

Important information may be hidden from view in most drawings of the building, so it is helpful to be able to look at a 'slice' through the building to see the hidden detail. For example, on a working drawing for a house, a section (or sectional) drawing could allow us to see clearly the layout of the stairs within the building.

ACTIVITY

Imagine a 'slice' (a section drawing) from front to back through a house with two floors. Think of the features that could be shown besides a staircase. Make a list and discuss your ideas with a partner.

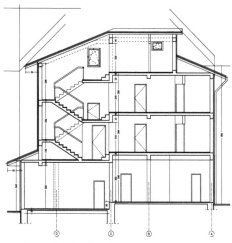

▲ Figure 1.6 Section drawing

Section drawings are produced to a scale of 1:50 or 1:100. These scales match those used for floor plans and elevations (discussed in later chapters), which makes it easier for the bricklayer to visualise and tie together a great deal of potentially complex information. Having as complete an understanding as possible of all the details of a structure is important to efficient planning and preparation for construction tasks.

Detail drawings

Detail drawings show large-scale details of the construction of a particular item, such as a timber-frame structural corner. (You can find more on timber-frame structures later in this chapter.) Another example would be a complex brickwork detail or feature.

Commonly used scales for detail drawings are 1:5 and 1:10.

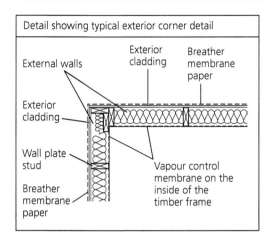

▲ Figure 1.7 Detail drawing

Symbols used in construction drawings

To identify building features, components and materials, a range of **standardised** symbols are used in construction drawings as a means of passing on information in a simple way. Using written labels to identify items such as electrical sockets or toilets and sinks would quickly clutter a drawing with text and could be confusing.

Types of materials can be represented on a section drawing by lines and symbols called **hatchings**. It is important that these conform to a standard, so that everyone using the drawing can interpret the information consistently and accurately. Study Figure 1.8 to see what symbols on a construction drawing mean.

KEY TERMS

Conventions: an agreed set of standards, practices and methods for producing drawings

Standardised: conforming to a set standard

Hatchings: a standardised set of lines and symbols that allow easy identification of materials shown on a drawing

Sink	Sinktop	Wash basin	Bath	Shower tray
WC	Window	Door	Radiator	Lamp
Switch	Socket	North symbol	Sawn timber (unwrot)	Concrete
Insulation	Brickwork	Blockwork	Stonework	Earth (subsoil)
Cement screed	Damp proof course (DPC)/membrane	Hardcore	Hinging position of windows	Stairs up and down
Timber – softwood. Machined all round (wrot)	Timber – hardwood. Machined all round			

▲ Figure 1.8 Basic drawing symbols and hatchings

1.2 Documents

While drawings are an effective visual method of communicating a great deal of important information, a range of written documents can also be used to support construction activities. These documents are used to record clear and accurate information that is communicated between site personnel.

You will learn about many of these documents at Level 2, but this chapter concentrates on documents that provide detail about the elements of a building and the sequence of construction.

▲ Figure 1.9 A range of documents can be used to communicate important information on site

Specification

The **specification** is a contract document, which means it is legally binding. It is used in conjunction with drawings. Giving written instructions in a separate document highlights important information and helps to prevent the drawing getting cluttered with text.

A specification provides clear information to the bricklayer about the work task. For example, it could give details of:

- which materials to use
- correct working practices
- the quality standards that must be achieved.

For instance, a specification for the bricks to be used for a work task would include information such as:

- type
- appearance
- **compressive strength**
- water absorption.

TECHNICAL SPECIFICATION

Brick Type				New Products (higher specification)		
Brick Type	**HANDMADE**	**TRADITIONAL WIRECUT**	**RECLAIM & CHERWELL**			
Brick Codes				CHB	VGB (Wirecut)	Cotswold Collection
Appearance	Lightly creased genuine handmade	Extruded wirecut smooth, rustic, plain or sandfaced	Prematurely 'aged' during the production process			
Specification	Bricks are manufactured to BS EN 771-1 (BS 3921:1985 now withdrawn)					
Sizes	Metric: 215mm x 102.5mm x 50mm, 65mm 68mm 73mm 80mm Imperial: 9" x 4⁵⁄₁₆" x 2", 2¼", 2³⁄₈", 2½", 2⁵⁄₈", 2⁷⁄₈", 3", 3¹⁄₈"		Non Standard: Most sizes can be accommodated in our production process			65mm 73mm
Specials	A full range of standard BS and purpose made Specials is available					
Compressive Strength	>24 N/mm²	>60 N/mm²	>60 N/mm²	>60 N/mm²	>60 N/mm²	>60 N/mm²
Durability	F2	F2	F2	F2	F2	F2
Tolerances	T1	T2	T1	T1	T1	T1
Range	R1	R1	R1	R1	R1	R1
Soluble Salt Content	S2	S2	S2	S2	S2	S2
Water Absorption	<17 %	<12 %	<12 %	<12 %		
Packaging	Shrinkwrapped and wire banded with holes for forklift use					
Pack Sizes & Weights	Size (mm) / Qty/pack / Wt (t) 50 / 650 / 1.229 65 / 500 / 1.223 73 / 500 / 1.280 80 / 400 / 1.213	Size (mm) / Qty/pack / Wt (t) 50 / 650 / 1.070 65 / 500 / 1.072 73 / 500 / 1.195 80 / 400 / 1.100	Size (mm) / Qty/pack / Wt (t) 50 / 650 / 0.927 65 / 435 / 0.929 73 / 400 / 1.009 80 / 320		Size (mm) / Qty/pack / Wt (t) 50 / 650 / 1.070 65 / 500 / 1.072 73 / 500 / 1.195 80 / 400 / 1.100	
Workmanship	The recommendations of good practice made in the relevant British or European Standards regarding design and workmanship must be fully observed. Always mix from at least 3 packs, working diagonally down the blades, not horizontally across the tops of packs. Bricks & brickwork should be covered during construction to prevent saturation.					
Performance	All Northcot 1st quality bricks are Frost Resistant. Northcot Bricks are manufactured to F2 rating for Frost Resistance as described in BS EN 771-1: 2011 *"Specification for Clay masonry units"*. The specification is for masonry (walling) products which are to be used in brickwork designed and built in accordance with recommendations in BS PD 6697: 2010 *"Recommendations for the design of masonry structures to BS EN 1996-1-1 and BS EN 1996-2"*. The use of Sulphate Resisting Cement is recommended. Further details on website					
Samples	Usually within 48-hour dispatch					
Important Advice	Colours and textures reproduced here are as accurate as the printing process allows and final choices should not be made from this brochure in isolation. For the latest updates on technical information, please visit our website on **www.northcotbrick.co.uk**					

▲ Figure 1.10 Specification

A specification is usually a separate document that is viewed in conjunction with a drawing. However, a small specification panel is sometimes included along the side edge of a drawing, to provide information about important details such as:

- type of wall ties
- type of insulation
- overall dimensional width of a cavity wall.

This means the bricklayer will have a comprehensive source of information in a single document to take on site.

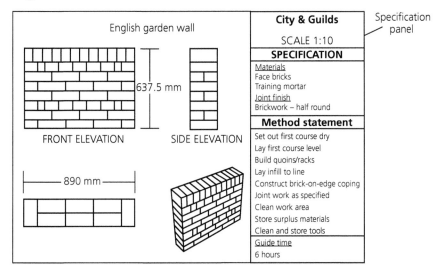

▲ Figure 1.11 Working drawing of a solid wall showing the specification panel

IMPROVE YOUR ENGLISH

Study the drawing specification panel in Figure 1.11. Compare the written content in the specification and the method statement. Write a short explanation of the differences.

Schedules

Larger, more complex construction sites (for example, a housing site where several house designs are being constructed) might have different components and fittings for each type of building. For example, there could be different colours of roof tile or different types of brick.

KEY TERM

Schedule: a list of repeating components or features showing the building or site location where they are intended to be installed

These details are listed in a **schedule**, which is used alongside corresponding drawings. Typical working drawings will have components such as doors and windows labelled D1, D2, W1, W2 and so on. These labelled components are listed in the schedule along with other information. This makes it easier to interpret repetitive data and reduces the chance of error.

Master Internal Door Schedule							
Ref:	Door size:	S.O. width	S.O. height	Lintel type	FD30	Self closing	Floor level
D1	838 × 1981	900	2040	BOX	Yes	Yes	GROUND FLOOR
D2	838 × 1981	900	2040	BOX	Yes	Yes	GROUND FLOOR
D3	762 × 1981	824	2040	BOX	No	No	GROUND FLOOR
D4	838 × 1981	900	2040	N/A	Yes	No	GROUND FLOOR
D5	838 × 1981	900	2040	BOX	Yes	Yes	GROUND FLOOR
D6	762 × 1981	824	2040	BOX	Yes	Yes	FIRST FLOOR
D7	762 × 1981	824	2040	BOX	Yes	Yes	FIRST FLOOR
D8	762 × 1981	824	2040	N/A	Yes	No	FIRST FLOOR
D9	762 × 1981	824	2040	BOX	Yes	Yes	FIRST FLOOR
D10	762 × 1981	824	2040	N/A	No	No	FIRST FLOOR
D11	686 × 1981	748	2040	N/A	Yes	No	SECOND FLOOR
D12	762 × 1981	824	2040	BOX	Yes	Yes	SECOND FLOOR
D13	762 × 1981	824	2040	100 HD BOX	Yes	Yes	SECOND FLOOR
D14	686 × 1981	748	2040	N/A	No	No	SECOND FLOOR

▲ Figure 1.12 Schedule

Programme of work

Careful planning is vital to achieve efficiency and productivity. A programme of work is used to plan the sequence of work tasks on site and often takes the form of a bar chart. Each company is likely to have its own variation, but traditionally this bar chart is referred to as a Gantt chart (as developed by Henry Gantt in the early twentieth century).

A Gantt chart is used to plan, monitor and understand a set sequence of activities and shows if work on site is progressing to schedule. It offers a clear view of overlapping activities and helps to identify when labour and equipment requirements need to be met.

Note that the section for each task in the sequence is split into two rows. The top row shows the planned activity timings and the bottom row is a different colour to show actual activity timings. Experience gained from past projects, coupled with information provided by **method statements**, can be used to work out how much time should be allotted to individual tasks.

KEY TERM

Method statements: documents giving information on safe and efficient sequences of work

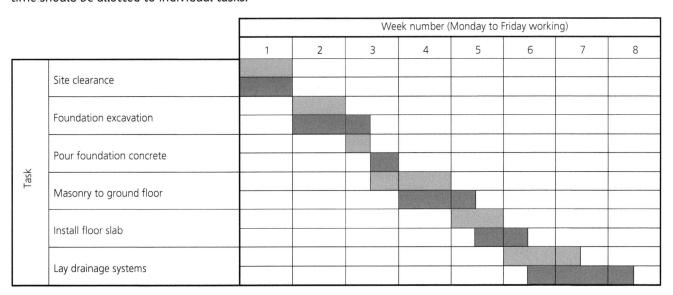

| | | Week number (Monday to Friday working) | | | | | | | |
		1	2	3	4	5	6	7	8
Task	Site clearance								
	Foundation excavation								
	Pour foundation concrete								
	Masonry to ground floor								
	Install floor slab								
	Lay drainage systems								

▢ Planned activity timings

▢ Actual activity timings

▲ Figure 1.13 Bar chart or Gantt chart

IMPROVE YOUR MATHS

Look carefully at the Gantt chart in Figure 1.13. Although it is set out to show the number of weeks planned for each task, work out how many days were spent on the following activities:

- foundation excavation
- masonry to ground floor
- installation of floor slab.

2 HOW CONSTRUCTION CAN AFFECT THE ENVIRONMENT

For some time, there has been concern about how construction activities can impact on the **environment**. A lot of effort is being put into making both construction methods and completed buildings more environmentally friendly.

Constructing and operating buildings uses large amounts of energy. That is why it is important to make greater use of renewable energy sources, such as solar and wind, both in the manufacture of construction materials and components and in the way buildings are designed to operate.

▲ Figure 1.14 Solar panels

▲ Figure 1.15 Wind turbines

2.1 Saving energy and water

Building regulations play an important part in controlling construction activities to support **sustainability**.

For example, the regulations state the minimum levels of insulation required in buildings to control heat transfer:

- Insulation needs to reduce heat loss from a building during cold weather. This in turn reduces the energy needed to maintain a comfortable temperature for the occupants.
- Insulation also needs to reduce the heat entering a building in hot weather. This saves energy by reducing the need for air-conditioning systems.

KEY TERM

Sustainability: meeting our own needs without damaging the ability of future generations to meet their own needs

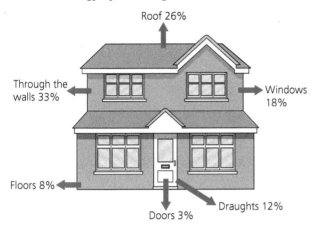

▲ Figure 1.16 Sources of heat loss from a house

▲ Figure 1.17 Air-conditioning unit

New structures may be designed to include drainage systems that store rainwater in tanks, often buried underground to allow heavy storage tanks to be used. This is called water harvesting, and the water stored can be used to flush toilets through what is known as a 'grey water' system. This reduces the consumption of treated water resources and saves energy.

▲ Figure 1.18 Rainwater stored in an underground tank

ACTIVITY

Search online for 'energy-saving measures in construction'. Identify three ways that energy can be saved either during the construction process or in the completed building.

2.2 Choosing and sourcing materials

Energy used in manufacturing and transporting construction materials creates what is referred to as a '**carbon footprint**'.

For example, a lot of energy is used when manufacturing concrete and transporting it to site. Traditional methods of energy production (such as burning fossil fuels) generate carbon, which combines with oxygen to produce carbon dioxide gas in the air. For this reason, many producers use recycled **aggregate** in the concrete mix, to reduce its carbon footprint.

KEY TERMS

Carbon footprint: the amount of greenhouse gases – primarily carbon dioxide – released into the atmosphere by a particular human activity

Aggregate: mineral material in the form of grains or particles, typically made of sand, stone, gravel, recycled concrete and crushed rock

▲ Figure 1.19 Concrete production plant

It is important to choose materials that have the lowest carbon footprint possible. For example, materials obtained from nearby sources will not have been transported over long distances; this reduces the amount of carbon produced.

The choice of construction method is especially important. For example, making greater use of timber construction methods can reduce damage to the environment, as long as the materials are sourced from responsibly managed forests, where trees are replanted after harvesting to provide a sustainable source of timber.

2.3 Managing waste

Another important environmental consideration is reducing the amount of waste created during both the manufacture of materials and construction activities. For example, to make manufacturing processes more sustainable, some insulating materials are now made from recycled glass and plastic which previously was disposed of as waste.

▲ Figure 1.20 Fibreglass insulation can be manufactured from recycled glass

When old buildings are demolished, many of the construction materials can be reused or recycled. Concrete paths, roadways and floor slabs can be crushed and recycled as aggregate for new concrete or **hardcore**. Bricks can be reclaimed and reused or crushed to form aggregate or solid fill. Timber can be recycled in various forms.

KEY TERM

Hardcore: solid materials used to create a base for load-bearing concrete floors, paths or roadways

ACTIVITY

Search online for 'recycling timber in construction'. List four things that recycled timber can be used for.

▲ Figure 1.21 Concrete can be crushed and recycled

Unavoidable waste on site should be **segregated** into specified groups or categories of materials, such as wood, glass, rubble and metal, to make off-site recycling easier.

▲ Figure 1.22 Segregated waste skips on site

3 THE CONSTRUCTION OF DIFFERENT PARTS OF BUILDINGS

All buildings have certain parts or elements in common, namely the foundation, floors, walls and roof. In this section, we will look at each element in turn and consider the principles that govern their construction.

3.1 Foundation

A foundation supports the building and transfers the weight or 'loadings' of the structure to the natural foundation, which is the ground on which it sits. A foundation is part of the design known as the **substructure** and is the main **load-bearing** part of a building.

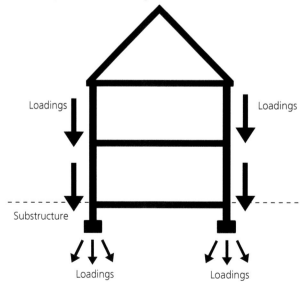

▲ Figure 1.23 Loadings are transferred to the ground through the foundation

3.2 Types of foundation

Most foundations are formed in concrete. Concrete is made from fine aggregate (sand) and coarse aggregate (crushed stone) mixed with cement and water.

Sometimes a concrete foundation will require **reinforcement**, perhaps to span a section of softer ground or to carry higher loadings created by a larger building. This is achieved by placing steel bars or mesh into the concrete in a specified pattern before it sets hard.

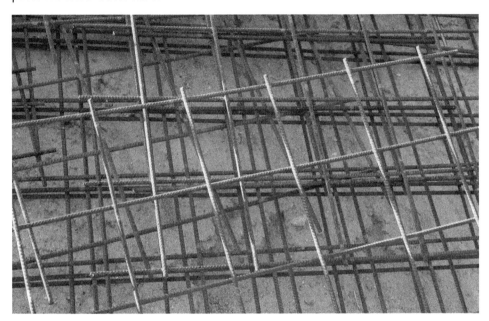

▲ Figure 1.24 Steel reinforcing mesh

Many factors must be considered when determining the type of foundation required and the design dimensions needed to carry the loadings of the building structure:

- soil conditions
- depth of bedrock
- depth of water table
- effect of nearby trees
- sloping ground
- ground contamination
- size and weight of the building.

After careful consideration of the relevant details, the foundation design will be finalised and the requirements written into a specification. This will include details of the:

- type of concrete to be used
- type of steel reinforcement if appropriate
- methods of installation where relevant.

The specification can be linked with the detail and section drawings to show the placement of steel reinforcement and the dimensions of the concrete foundation.

KEY TERMS

Volume: the total space in three dimensions taken up by an object, material or substance

Datum: a reliable fixed point or height from which reference levels can be taken

Friction: the resistance that one surface or object encounters when moving over another

These details make it possible to calculate the amount of materials required. The quantities of coarse aggregate, sand and cement are calculated according to the specified ratio, and the design dimensions are used to calculate the **volume** of concrete needed. (There is an explanation of how to calculate volume later in this chapter, along with some examples to try.)

IMPROVE YOUR MATHS

Volume is calculated by multiplying length by width by depth. Because three numbers are multiplied together, the result is expressed as a 'cubic' measurement like this: m^3.

Different foundation designs may require different ratios of material quantities, to vary the strength of the concrete. In some cases, special chemicals are added to the concrete mix, to either speed up the hardening process (accelerators) or slow it down (retarders). Other additives can be used to waterproof the concrete or protect it from frost.

ACTIVITY

Think about situations where the hardening process of concrete might need to be speeded up or slowed down. Make a list and discuss your ideas with a partner or carry out some research about the subject.

Table 1.1 shows different types of foundation design and how they can be used.

▼ Table 1.1 Types of foundation

Type of foundation	How it can be used
Strip 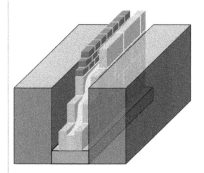 	Strip foundations are widely used for housing and small commercial developments. The design specification will state the width and depth of the trench that must be excavated to suit the soil conditions and the weight of the building. A strip of concrete is poured into the excavated trench. The thickness of the strip will depend on the design. This will be of a minimum thickness of 150 mm. More commonly the thickness will be increased to 225 mm. The surface of the concrete should be levelled carefully to allow the bricklayer to start work on a uniform level surface when constructing the foundation masonry. The foundation level can be set by measurement from **datum** pegs set at the corners of the excavation. (See Chapter 2 for more information on datum points.)

▼ Table 1.1 Types of foundation (continued)

Type of foundation	How it can be used
Trench fill	If the full depth of the excavated trench is filled with concrete, this is referred to as 'trench fill'. This type of foundation may be used where there are trees close by, because tree roots can undermine a foundation and they also extract water from the ground, causing the soil to shrink. This can lead to failure of the foundation, which will severely affect the structure of the building.
Pile	Piles are essentially long cylinders of a strong material such as steel or concrete. They are used to transfer the load of a building through soft or unsuitable soil layers into the harder layers of ground below, even down to rock if required. This type of pile foundation is known as 'end bearing' and is effective when a building has very heavy, concentrated loads, such as in a high-rise structure or a bridge. A second type of pile foundation is known as 'friction pile'. Support for a building is provided by the full height of the pile creating **friction** with the soil it stands in. The deeper into the ground the pile is driven, the greater the friction and load-bearing capacity. Installing piles requires specialist equipment and trained personnel and can be an expensive option.
Raft	A raft foundation is often used where a strip foundation would be unsuitable due to soft ground conditions or where a pile foundation would be too expensive. It consists of a reinforced slab of concrete covering the entire base of the building, spreading the weight over a wide area. The edge of the slab is usually thickened as a support for load-bearing walls around the outline of a building. If any minor movement takes place due to poor ground conditions, the building is protected, since the whole foundation can move slightly as a unit.
Pad	A brick pier or a structural steel column in a steel-framed building will produce loadings concentrated on a single point. A pad foundation can be designed with greater depth and additional reinforcement to support this type of load. When a pad foundation is used to support steel framing, it can have bolts cast into the top, allowing steel columns to be fixed to it.

ACTIVITY

Carefully read the descriptions of different foundation types in Table 1.1. Then work with another learner to decide which type of foundation would be suitable for a single detached garage to be built on a plot with areas of soft ground near a group of trees. Write down the reasons for your choice.

3.3 Floors

Floors are load-bearing elements of a building, carrying the weight of occupants, equipment and furniture. The loadings they carry can be transferred through the walls on which they rest or are attached to, down to the foundation and ultimately to the ground on which the building sits.

There are various types of floor, which are constructed using different methods. The design depends on the type of building and the potential load that the floor will be required to carry.

Ground floors

Ground floors can be constructed as a solid or suspended design using a range of materials.

Solid concrete ground floors

A solid concrete floor design is often referred to as a 'ground-bearing' floor, because the weight of the concrete slab and loads placed on it are transferred to the ground directly below.

The concrete slab of a solid floor is laid on **compacted** hardcore. A damp-proof membrane (DPM) is installed under the slab, to prevent damp rising through the floor. The DPM is a sheet of strong waterproof material, usually laid on sand referred to as 'blinding' to prevent the hardcore puncturing it. The thickness of the DPM is referred to as its gauge, for example, '1000 gauge'.

Insulation can be installed either under or over the floor depending on the design, to reduce heat transfer. Some solid concrete floors are produced with a smooth surface, using specialist finishing equipment as the concrete hardens. If the surface of the slab is left with a rough surface, it can be given a smooth finish later by laying a thin layer of sand and cement called a 'screed'.

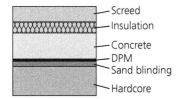

Screed
Insulation
Concrete
DPM
Sand blinding
Hardcore

▲ Figure 1.25 Section view of a concrete floor

KEY TERM

Compacted: firmly packed or pressed together

INDUSTRY TIP

You may hear the concrete slab used for a solid ground floor referred to as the 'oversite' concrete.

▲ Figure 1.26 Providing a smooth finish to a solid concrete floor

▲ Figure 1.27 Laying a screed finish to a solid concrete floor

Suspended timber ground floors

A traditional method of constructing ground floors is to use timber beams called **joists**, which span the outer walls of a structure.

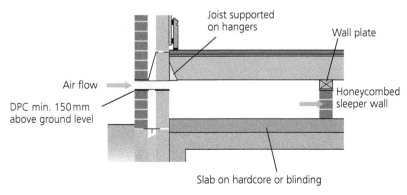

▲ Figure 1.28 Suspended timber floor

KEY TERM

Joists: parallel timber beams spanning the walls of a structure to support a floor or ceiling

The joists rest on lengths of timber referred to as 'wall plates', which in turn sit on supporting brick 'sleeper' walls built at intervals. The spaces between the joists can be used to install insulation. The sleeper walls are built on oversite concrete, which is laid on hardcore or, in some cases in the past, directly on the earth below the floor.

The gap between the oversite concrete and the joists is ventilated to prevent condensation, and a **damp-proof course** (DPC) is installed on top of the sleeper walls under the wall plate to protect the timber from damp. (There is more on DPC in Chapters 4 and 5.)

Timber floor boarding, or timber sheet material such as a suitable grade of chipboard, is fixed across the joists to form the floor surface.

Upper floors

Suspended timber upper floors

Upper floors are all suspended in that the supporting joists are attached to or bear on the walls of a structure. There are several ways of linking the timber joist to the walls to produce a solid and stable result. Steel connectors and hangers in a range of shapes and sizes allow quick and easy installation of timber beams when constructing upper floors.

KEY TERM

Damp-proof course: a continuous barrier built into masonry at specified locations to prevent moisture entering a structure

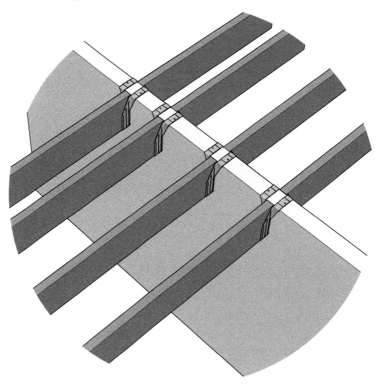

▲ Figure 1.29 Timber joists supported with steel hangers

ACTIVITY

There are suspended ground- and upper-floor designs that are constructed in concrete materials. Research suspended concrete floors and produce a sketch of what they look like.

3.4 Walls

A bricklayer will build two main types of masonry wall: solid walls and cavity walls.

Solid walls vary greatly in thickness and can be constructed using bricks or blocks or a combination of both. An example is a thick solid wall referred to as a 'retaining' wall, which can be used to hold back a mass of earth or other material.

Cavity walls are generally constructed in two single leaves or skins of masonry, separated by a cavity of specified width. Like solid walls, they can be built using bricks or blocks or a combination of both.

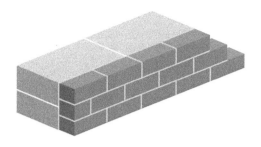

▲ Figure 1.30 Retaining wall in brick and block

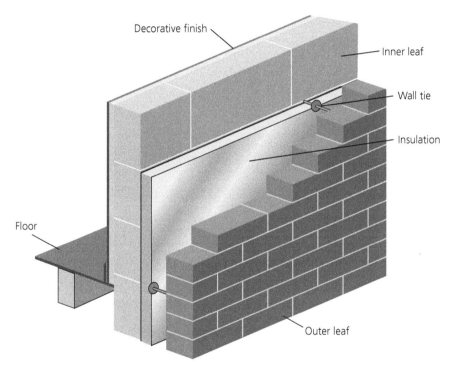

▲ Figure 1.31 Cavity wall

Chapters 3 and 4 cover single-leaf, half-brick-thick solid walls constructed using blocks and bricks. Solid walls constructed to a greater thickness are discussed in more detail at Level 2. Chapter 5 covers cavity wall construction in detail.

This chapter considers how external and internal walls are constructed using materials other than masonry.

Types of wall

Timber-frame external walls

Modern timber-frame houses are precision engineered, strong and durable. They are commonly constructed from factory-made panels which are transported to site ready for assembly. The modern arrangement is to have a timber frame that carries the weight of the floors and roof and an outer leaf (or skin) of masonry or **cladding**, such as treated timber, to waterproof the structure and provide an attractive appearance.

KEY TERM

Cladding: a covering or coating of one material over another to provide a skin or layer

21

ACTIVITY

Search online for 'cladding'. Write a short description of as many different types of cladding as you can find.

▲ Figure 1.32 A timber-frame structure

The timber frame and outer leaf are tied together with wall ties, and the structure is protected from damp by the installation of a DPC at specific locations. (See Chapter 5 for more on using wall ties and DPC in cavity walls.)

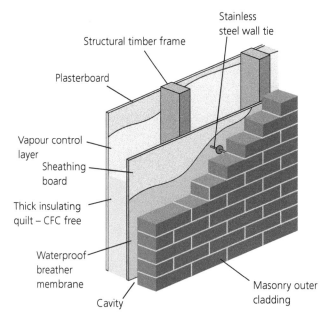

Stainless
steel wall tie

Structural timber frame

Plasterboard

Vapour control
layer

Sheathing
board

Thick insulating
quilt – CFC free

Waterproof
breather
membrane

Cavity

Masonry outer
cladding

▲ Figure 1.33 Timber-frame cavity wall

INDUSTRY TIP

The bricklayer plays an important role in making sure the timber-frame structure performs as it should. Wall ties must be installed so as not to damage any moisture barriers, fire stops must be fitted as specified, and mortar must not be allowed to build up against the timber core or where the DPC is positioned.

Using a timber-frame structure has many advantages:

- It can speed up the building process, since the internal timber frame can be quickly erected and temporarily weatherproofed, and internal work can start at the same time as the masonry outer leaf or external cladding is being constructed.
- It can be highly insulated to reduce heat and sound transfer.
- If it uses resources from responsibly managed forests, it is a sustainable building method using renewable materials.

Internal walls

Walls within a structure can be:

- load-bearing if they support floors above them, or
- non-load-bearing if used simply as partitions to divide large internal spaces.

They can be **prefabricated** or assembled on site at the work location.

Internal walls can be constructed using timber or metal as a framework, ready to be covered in a material such as plasterboard or another suitable sheet material. Wall elements produced in this way are commonly referred to as 'stud walls' or 'studding'.

ACTIVITY

Search online for a manufacturer of timber-frame buildings. Choose a design from its website, copy and paste an image of your choice into a Word document and write a short explanation of why you chose it.

KEY TERM

Prefabricated: factory-made units or components transported to site for easy assembly

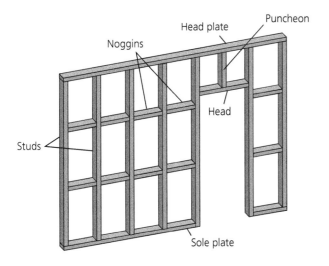

▲ Figure 1.34 Timber stud wall

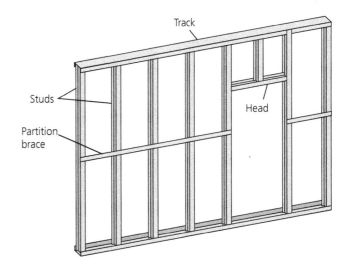

▲ Figure 1.35 Metal stud wall

When used for load-bearing situations, internal walls are usually constructed using blocks manufactured from dense concrete. Lightweight concrete blocks can be used, but this depends on the load-bearing requirements stated in the specification.

Wall finishes

A face-brick wall, when completed, provides an external finish that is durable and long lasting, or if used internally can remain an attractive feature for many years. Other types of wall construction such as blockwork may require additional materials to be applied to provide a suitable finish.

External finish

To provide a finish for external walls, a sand and cement mix known as render can be applied. The mix ratio is carefully formulated to allow for any expansion and contraction that may occur within the surface over which the render is applied.

The mix may include powdered lime or a chemical **plasticiser**, to make the render more flexible and workable. (There is more on plasticisers in Chapter 3.) The addition of lime powder can also make the render mix stickier, for easier application on a vertical surface.

A skilled operative can produce a smooth, long-lasting render finish, usually applied in two or sometimes three coats. The surface can then be painted with a masonry paint if required. Other types of external finish are covered at Level 2.

> **KEY TERM**
>
> **Plasticiser:** an additive used to make mortar more workable and pliable

▲ Figure 1.36 Applying a coat of render to an external wall

> **INDUSTRY TIP**
>
> When you build a wall that will be rendered later, make sure that the mortar joints are finished flush and not left projecting from the wall. The plasterer will thank you for leaving a smooth surface to work on!

Internal finish

Internal walls (and ceilings) can be plastered, which means they are covered with two or more thin coats of plaster over plasterboard. This gives a smooth surface which is then usually finished with emulsion paint or papered coverings.

The finish provided by emulsion paint will be more durable on new plaster if the surface is first sealed with a diluted coat of paint called a mist coat, followed by two undiluted coats.

▲ Figure 1.37 Plaster gives a smooth finish

3.5 Roofs

Roofs protect the structure they sit above by providing a weatherproof surface that directs rain to storm-water drainage systems. They must be strong enough to withstand high winds and the potential weight (or loading) of snow standing on them.

Because heat rises, a large amount of heat energy can be lost through the roof as the uppermost part of the building. It is therefore important to install adequate thermal insulation in any type of roof.

Types of roof
Flat roofs

A flat roof is not literally flat. It must have a slope or incline of up to 10° to prevent rainwater building up on the surface. Traditionally, the waterproof coating of a flat roof comprised of felt material covered in tar (called bituminous), built up in several layers. An improved, longer-lasting covering is provided by layers of glass-fibre sheets impregnated with a special resin.

ACTIVITY

Wind is powerful enough to cause significant damage to a roof. Search online for images of wind-damaged roofs and note the extent of the damage that can occur.

▲ Figure 1.38 Flat roof

Pitched roofs

A pitched roof has a weatherproof surface or surfaces that slope at more than 10°. They are constructed using timber components called **rafters** and have a range of different design shapes. (A horizontal beam is a 'joist', as previously mentioned in this chapter.)

A pitched roof can be constructed with a single sloping surface that leans against an adjoining wall, appropriately called a 'lean-to'. This type of roof is commonly used for porches and extensions to the main building. If a single pitched roof surface covers the entire building, this is referred to as a 'mono-pitch' roof.

▲ Figure 1.39 Lean-to roof

Where a roof has two pitched or sloping surfaces, with triangular walls closing each end of the roof up to the **ridge** (or apex), this is referred to as a 'gabled' roof. The triangular walls are known as gables or gable ends.

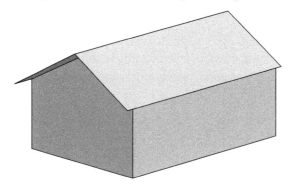

▲ Figure 1.40 Gabled roof

KEY TERMS

Rafters: beams set to suit the angle of the roof pitch, forming part of the roof's internal framework

Ridge: the highest horizontal line on a pitched roof where the sloping surfaces meet

A 'hipped' roof has no gables, and the weatherproof surfaces slope down from the ridge to the top of the walls on all sides. These roofs are more complex, and therefore more expensive, to construct.

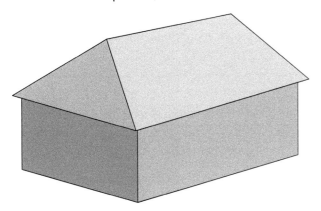

▲ Figure 1.41 Hipped roof

Pitched roofs are often designed to project beyond the face of the wall they sit above, to help protect the wall from bad weather. Look carefully at this design feature as shown in Figure 1.42.

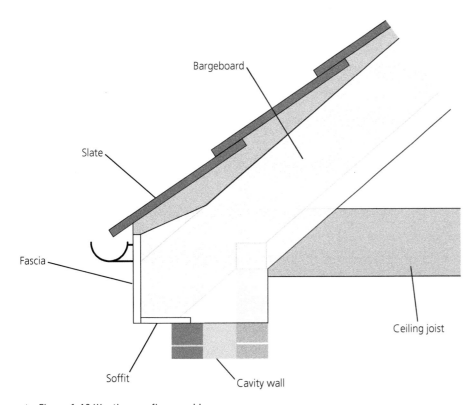

▲ Figure 1.42 Weatherproofing a gable

If timber is used for exposed components like the facia, it will need to be treated and painted to preserve it over time. Typically, the knots in the timber will be treated (called 'knotting') and the wood will be painted with a primer, undercoat and gloss finish to give the maximum protection.

Many exposed components traditionally produced in timber are now manufactured in plastic which does not rot and reduces the need for painting.

Roof components

Traditionally, roofs were constructed on site from individual lengths of timber. These are referred to as 'cut' roofs, since the timber is cut to the required lengths and angles to suit the design. This is a process that requires carpentry skills and a good understanding of geometry.

Study the hipped cut roof in Figure 1.43 to become familiar with the terminology used for roof components.

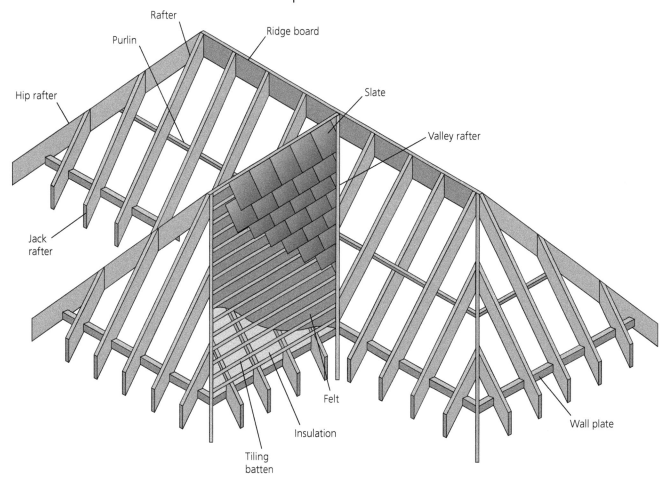

▲ Figure 1.43 Roof components

Most roofs are now constructed using 'trussed' rafters. This system uses factory-made timber components that are delivered to site and assembled more quickly than a cut roof, thereby reducing costs. Using this engineered system means that lighter timbers can be used, making additional cost savings.

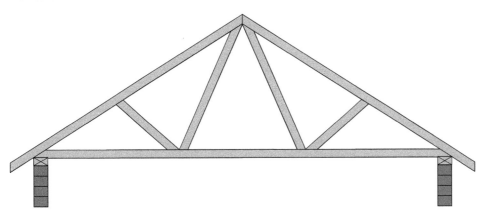

▲ Figure 1.44 Trussed rafter

Some roofs require a combination of cut-roof methods and trussed rafters. All roofs must be securely anchored to the walls of the building to conform to building regulations. This is achieved by using mild-steel restraining straps at specified positions in the structure.

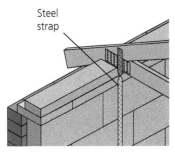

▲ Figure 1.45 Restraining strap

Roof covering materials

Once the roof structure is complete, a surface covering is installed which must be highly water resistant and durable. It is usually fitted over a felt or **semi-permeable** membrane or sheet material that protects against windblown rain or snow entering the roof space.

Traditionally, slate has been favoured as a roof covering. This is a natural product which can be split into thin sheets by skilled workers. Slate continues to be used extensively as a roof covering, despite its relative expense.

KEY TERM

Semi-permeable: material that allows only certain materials or substances (such as water vapour) to pass through it

▲ Figure 1.46 Natural slate roof covering

An alternative to slate is tiles:

- Clay tiles are manufactured from a natural material and are produced in a range of sizes and shapes to suit different applications. They produce a finished roof covering that is attractive and durable.
- Concrete tiles are manufactured in different sizes, shapes, textures and colours. They are relatively quick to install and are often designed to interlock with each other to form a wind-resistant covering.

▲ Figure 1.47 Concrete roof tiles

Both clay tiles and concrete tiles are heavy components, requiring a roof structure that can carry the loadings adequately.

ACTIVITY

Research the following roof coverings: shingles, thatch and turf. Write a description for each of the materials and describe how they are installed.

4 CALCULATING QUANTITIES OF MATERIALS

Now that we are familiar with the elements of a building and the materials from which they are constructed, we will look at methods used to calculate the quantities required.

Accurate calculation of quantities of materials is important to support efficiency and productivity, as well as to minimise waste.

4.1 Volume

The volume of an object or substance is the total space it takes up in three dimensions, for example, the volume of concrete required for a concrete foundation or the drum capacity of a concrete mixer.

▲ Figure 1.48 The amount of concrete required is calculated as volume

As mentioned previously, to calculate the volume of a space or object, we multiply the length by the width by the height to give the volume expressed in cubic metres (m^3):

volume = length × width × height (L × W × H)

IMPROVE YOUR MATHS

1 A brick wall needs a foundation with the following dimensions: length 4 m, width 0.6 m and height 0.3 m.

Calculate the volume of concrete needed.

2 A block wall needs a trench-fill foundation with the following dimensions: length 6.5 m, width 0.4 m and height 1.2 m.

Calculate the volume of concrete needed.

4.2 Area

Area is how we describe the measurement of a two-dimensional surface, such as the face of a wall. To find the surface area of the face of a wall, multiply its length by its height:

area = length × height (L × H)

The result is expressed in square metres (m^2).

Knowing the area of the face of a wall in m² allows easy calculation of the quantities of bricks or blocks needed for a work task. For a half-brick-thick wall, simply multiply the number of m² by 60 for bricks or by 10 for blocks.

IMPROVE YOUR MATHS

1 A brick wall will be built to the following dimensions: length 5.5 m and height 0.9 m.

  Calculate the area of the wall and the number of bricks needed.

2 A block wall will be built to the following dimensions: length 9 m and height 0.45 m.

  Calculate the area of the wall and the number of blocks needed.

4.3 Linear measurement

Linear measurement is the distance measured along a line between two points. You will use linear measurements to find, for example, the number of drainage pipes required or the length of facia board needed for a roof structure.

In a rectangular building, by adding up the separate linear measurements of the four sides we can establish what is known as the **perimeter**.

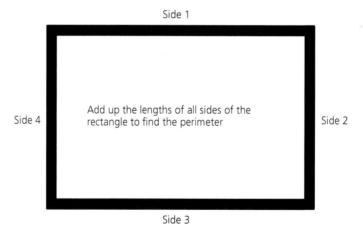

▲ Figure 1.49 Finding the perimeter

IMPROVE YOUR MATHS

1 A flat roof for a rectangular garage will be built to the following dimensions: length 6 m and width 5.5 m.

  To establish how much facia will be needed in linear metres, calculate the perimeter measurement.

2 The linear measurement that a rainwater drain extends from the garage to the plot boundary is 5.4 m. The drain is made up of separate pipe units, each 0.6 m long.

  How many pipe units will be needed?

5 COMMUNICATING IN THE WORKPLACE

Effective communication skills are crucial in developing and maintaining good working relationships and in achieving success in the workplace.

5.1 Methods of communicating

Verbal

Every day we naturally use different communication methods without thinking about it. We might use **verbal communication** when talking to others face to face or on a telephone, or sometimes using a two-way radio on site.

▲ Figure 1.50 Verbal communication can occur in many ways

Keep in mind that mistakes can easily be made while communicating verbally. The person giving the information might not make the matter clear enough, or the person receiving the information might misunderstand something. Often on site there will be a lot of background noise, which can lead to information being misheard.

Written

There are many forms of written communication. In the past, a 'memo' was commonly used – a brief note as a reminder of something that needs to be done. Memos have mostly been replaced by emails, which can be used to provide written reminders, ask questions or pass on instructions.

Text messaging can be used to exchange information quickly and conveniently when on site.

KEY TERM

Verbal communication: transmitting information by talking to others

IMPROVE YOUR ENGLISH

Memo is short for 'memorandum', which is Latin for 'to be remembered'. Many words used every day in English have Latin origins.

▲ Figure 1.51 Written communication can take place by email

Written communication can form a permanent record of information passed to others. If it is read and interpreted carefully, it can reduce the chance of misunderstanding.

Sometimes, a written record of a verbal conversation may be needed, for example, important instructions that may have been passed on in a phone call. When making a written record of a conversation that others may need to read, you should include:

- the date (and possibly time) that the conversation took place
- a brief but clear account of the content of the message
- the name and contact details of the person who passed on the information.

ACTIVITY

Working with a partner, take turns to imagine a message that you want to give to the other person that must be written down.

The message could be about a delivery of materials or a worker who has not arrived for their shift, or anything else you can think of related to working on site. Remember to include all the points listed above in the written message.

5.2 Benefits of effective communication

Poor communication can result in misunderstandings and mistakes, which in turn can lead to wasted materials, time and effort. This can have a serious effect on productivity. Furthermore, poor communication can also lead to accidents.

Keep in mind that good communication includes showing respect for others. The way personnel communicate with one another can affect motivation and morale on site, and respectful communication is essential if you are to play your part in supporting equality and diversity in the workplace.

Showing respect and consideration when dealing with colleagues can also influence the reputation a company builds with its customers over time.

▲ Figure 1.52 Maintain a respectful attitude when communicating with colleagues

A clear understanding of roles and responsibilities on a construction site will help you to communicate in an appropriate and effective way.

5.3 Job roles on site

Personnel in a construction team can be categorised according to the type of work they carry out, for example:

- professionals
- technicians
- trade operatives
- general operatives.

Each team member has an important part in applying the principles of construction discussed in this chapter.

Professionals

Professionals are trained and qualified to perform specific tasks. The training may take the form of many years of study to gain a recognised qualification. Table 1.2 outlines some professional roles.

▼ Table 1.2 Professional roles in construction

Role	Description
Architect 	Creates the concept and design of a building in accordance with the requirements of a **client**
Quantity surveyor 	Calculates quantities of materials, time needed for work tasks and labour costs
Surveyor 	Makes exact measurements and determines property boundaries; calculates heights and depths
Civil engineer	Plans, designs and oversees construction of large projects such as roads, railways and bridges

Technicians

Technicians provide support for professionals. They are trained to understand the technical aspects of whichever job they take on. For example, an architectural technician assists an architect in producing drawings for a project, which frees the architect to concentrate on concept and design matters and dealing with the client.

Trade operatives

A range of skills are needed for a construction project to be successful. Trade operatives form the 'backbone' of the construction industry. As a bricklayer, you are continuing a long history of skilled workers who have made a significant contribution to the built environment we see around us today.

General operatives

General operatives perform a range of semi-skilled and non-skilled work. They might perform essential tasks such as driving plant or machinery, working with the bricklayers mixing mortar or looking after site storage facilities.

▲ Figure 1.53 A general operative loading a mixer

Importantly, they may check deliveries to a site to make sure the correct quantities and types of materials are supplied undamaged and with no ordered items missing.

The smooth running of a construction site of any size is heavily dependent on all personnel being reliable and trustworthy and working cooperatively with colleagues.

Test your knowledge

1 Which term is used to describe the foundation, floors, walls and roof of a building?

   a Factors
   b Elements
   c Modules
   d Components

2 When a working drawing is legally binding, how is it referred to?

   a Contract document
   b Control document
   c Correct document
   d Command document

3 To correctly orientate a new building on a plot or site, which direction is shown on a site plan?

   a South
   b West
   c East
   d North

4 To which scale is a section drawing produced?

   a 1:5 or 1:10
   b 1:50 or 1:100
   c 1:100 or 1:200
   d 1:1250 or 1:2500

5 When insulation is specified in a building, what is it designed to control?

   a Cold transfer
   b Vibration transfer
   c Moisture transfer
   d Heat transfer

6 Which type of foundation would be used to carry loadings concentrated on a single point?

   a Pad
   b Raft
   c Strip
   d Pile

7 Above which angle is the slope of a pitched roof set?

   a 5°
   b 10°
   c 20°
   d 30°

8   Which roof component is a structural beam that follows the slope of a
    pitched roof?

    a   Facia

    b   Ridge

    c   Joist

    d   Rafter

9   When calculating quantities of drainage pipes for a project, which type of
    measurement is used?

    a   Area

    b   Linear

    c   Volume

    d   Percentage

10 Which category of personnel describes a quantity surveyor?

    a   Technician

    b   General operative

    c   Professional

    d   Trade operative

CHAPTER 2

SETTING OUT MASONRY STRUCTURES

INTRODUCTION

Setting out masonry structures requires great care in interpreting and applying the dimensions and measurements that form the design. Accuracy is vital at the setting-out stage of the construction process, to ensure the finished building is completed exactly as intended.

This chapter discusses:
- methods of preparing and clearing the site before starting the setting-out process
- equipment requirements
- methods used to set out and position a building on a site correctly
- processes used to set the height level of a structure in relation to specific reference points.

LEARNING OUTCOMES

After reading this chapter, you should:
1. know how the site is prepared for setting out
2. be able to identify tools and equipment used in setting out
3. know about the types of drawing used when setting out
4. know how to accurately set out a masonry structure to the set level.

---

1 PREPARING THE SITE FOR SETTING OUT

Good preparation at the setting-out stage is essential to prevent problems arising later in the project. Long before bricklayers arrive on site, a great deal of work will have been done to ensure the project proceeds safely and efficiently.

1.1 Services on site

Accidental damage to any existing services on site, such as electricity or gas, could pose a danger to workers, as well as causing delays and unexpected repair costs. Special equipment can be used to scan the ground to confirm the position of underground cables and pipe runs. It may be necessary to move the path of services.

▲ Figure 2.1 Underground services

As well as underground services, there may be overhead cables that must be considered during the preparation of the site. Overhead electricity cables can be a hazard to operators of machines such as cranes and excavators as they move about the site. Telecoms cables running underground or overhead must also be considered.

Water and drainage systems must be located and protected or rerouted. Accidental damage to water mains can cause disruption through flooding and may result in contamination of the water supply to nearby properties. Damage to drainage systems could cause pollution of the environment surrounding the site, so it is important that water courses and natural drainage ditches are protected during construction operations.

1.2 Site clearance

To prepare the site, **topsoil** must be removed from the area where buildings will be constructed. Topsoil is unsuitable to build on since it cannot support the weight (or loadings) a building will place on it. It also contains a large amount of soft vegetable matter.

If the topsoil is clean and of suitable quality, it can be stored for later reuse in landscaping and levelling the site when construction is finished.

INDUSTRY TIP

If topsoil is stored on site, the storage location must be considered carefully. The soil must be stored so that it does not interfere with water courses or drainage ditches.

Stockpiling in a location that will not mean moving the material again later is also an important part of efficient preparation of the site. Surplus topsoil of good quality is a valuable material and can be sold on.

▲ Figure 2.2 Natural drainage ditches must be protected

HEALTH AND SAFETY
Sometimes, a decision will be made not to remove or reroute services if they will not affect the positioning of buildings. If this is the case, control measures such as fences and signs must be put in place, to protect personnel from potential hazards.

KEY TERM

Topsoil: the upper, outermost layer of soil, usually the top 13–25 cm (5–10 inches)

▲ Figure 2.3 Removing topsoil

Removing topsoil creates a level area without obstacles, which makes it simpler to set out the outline of a building accurately. Even if the site has a slope, creating a uniform, smooth area will make setting out accurately much easier.

There may be existing buildings on a site, which must be demolished and removed. Creating a level area on such land presents additional challenges. The buildings that must be removed may have design features such as basements, or there may be other underground elements such as storage tanks. If so, the site will need additional preparation before new construction work can begin.

▲ Figure 2.4 Clearing a site of existing buildings

2 TOOLS AND EQUIPMENT USED IN SETTING OUT

The equipment used to set out a building may be relatively simple. Accurate results can be achieved using string lines attached to timber **profiles** as guides for setting out wall positions on the ground, ready for excavation of the foundations. These profiles can be left in place for later use by bricklayers when establishing wall positions on the completed foundation concrete.

KEY TERM

Profiles: (in the context of setting out) timber boards and pegs assembled at the corners and other wall locations of a building to allow string lines to be positioned accurately

INDUSTRY TIP

String lines used to set out a building are often referred to as ranging lines.

Profiles consist of timber rails (approximately 150 mm × 30 mm) attached to timber pegs (approximately 50 mm square) and are assembled on site to suit the requirements of the job. The pegs are securely driven into the ground and cross rails are fixed to them.

The positions of corner points and the lines of walls are indicated by nails driven into the top of the cross rails. An alternative method is to use saw cuts in the top of the rails to show the specified wall positions.

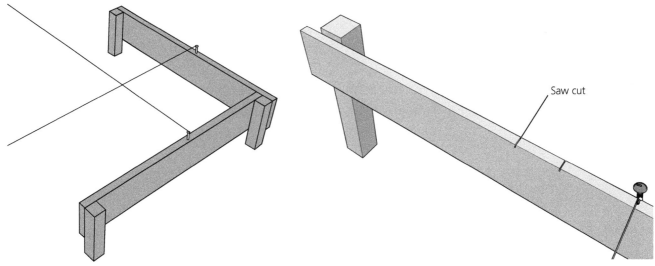

Saw cut

▲ Figure 2.5 String lines are attached to timber profiles   ▲ Figure 2.6 Saw cuts can be used to mark wall positions on timber profiles

INDUSTRY TIP

Using nails or saw cuts to mark wall positions on profiles is recommended, since the profiles could be in place for some time. Markings made in pen or pencil can wash away or fade, leading to mistakes in setting out.

Table 2.1 lists the tools and equipment used to set up the profiles and establish the building outline on the plot or construction site.

▼ Table 2.1 Tools and equipment for setting out wall positions

Tool/equipment	Description
Timber pegs (or stakes) and rails	These are used to construct profiles.
Lump (or club) hammer	This is used to drive pointed pegs firmly into the ground when constructing profiles and marking specific setting-out locations.
Carpenter's saw (hand saw)	This is used to cut profile rails to length. You may choose to use saw cuts on the top of the profile rails to indicate wall positions.
Tape measure/surveyor's tape measure	This is an important tool that is used constantly for measuring and marking dimensional details with accuracy. For small to medium measuring tasks, there is a range of lengths, from 3 m to 10 m. Longer tape measures are available.

▼ Table 2.1 Tools and equipment for setting out wall positions (continued)

Tool/equipment	Description
Mason's line and pins	Pins make it easy to attach string lines to reference points. They are effective over short to medium distances.
Ranging line	Over longer distances, a more substantial ranging line can be used. It is often made from nylon for strength.
Spray paint	This is used to mark guide lines on the ground, ready for excavation of the foundation trenches.

Additional items of equipment for creating right angles and transferring levels when setting out a building are discussed later in this chapter.

③ TYPES OF DRAWING USED WHEN SETTING OUT

In Chapter 1, we identified a range of drawings used to communicate information for a construction project. The use of scale was also considered, and Chapters 3 and 4 feature further application of scale when referring to working drawings. Drawing to scale means that the building being set out is represented with accurate proportions but reduced in size to fit onto a manageable size of paper.

In this chapter, we will concentrate on the drawings used to establish the position of a building on a plot or site, the building's outline and the position of internal walls.

3.1 Location drawings

Location drawings give information about the position of the new building in relation to existing roads, buildings and other features that are reliable fixed reference points to measure from. The position of a new building must comply with planning laws, so location drawings are an important source of information to make sure a building is constructed in the right place.

ACTIVITY

Search online for 'surveyor's tape measures' and make a list of all the different lengths of measuring tape you can find.

INDUSTRY TIP

Surveyor's tape measures are made from steel or special fabric. Fabric tapes should be used with care to avoid stretching them and distorting the measurements. Steel tapes should be clean and dry before winding them in after use, to avoid damage.

▼ Table 2.2 Types of location drawing

Type of drawing	Description
Block plan 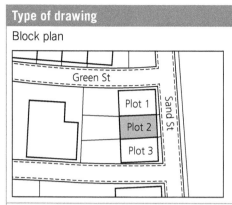	This is a 'bird's eye' view of the whole site in relation to the area around it. It shows individual plots and road layouts on the site as a simple outline with few dimensions. It allows planning of access requirements and can assist in planning storage facilities and positioning of materials during preparation for construction. Usual scale: 1:1250 or 1:2500
Site plan 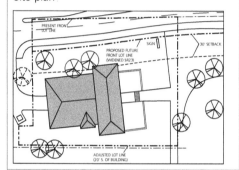	This shows the proposed development in relation to the site boundary, giving details needed to position a building accurately in accordance with local authority planning permission. It may show the position of drains on the site and could include the position of trees and shrubs if they are part of the planning requirements. Usual scale: 1:200 or 1:500

IMPROVE YOUR ENGLISH

Visit www.planningportal.co.uk. Go to the 'planning' section and write a short description of the purpose of the planning process.

3.2 Construction drawings

Construction drawings are used to understand the structural details of a building design. The design will be drawn in accordance with building regulations.

▼ Table 2.3 Types of construction drawing

Type of drawing	Description
Floor plan 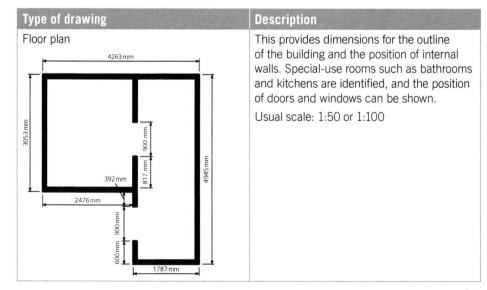	This provides dimensions for the outline of the building and the position of internal walls. Special-use rooms such as bathrooms and kitchens are identified, and the position of doors and windows can be shown. Usual scale: 1:50 or 1:100

▼ Table 2.3 Types of construction drawing (continued)

Type of drawing	Description
Section (or sectional) drawing	This is a slice through a structure which can show details that would otherwise be hidden. In the context of setting out a building, it could show the depth of a concrete foundation or other substructure details. Usual scale: 1:50 or 1:100
Detail drawing	This is a larger-scale drawing used to show complex design features in greater detail. Usual scale: 1:5 and 1:10

INDUSTRY TIP

Notice the lines on the brickwork and blockwork in the section drawing. These lines are used to identify the materials within the 'slice' of the wall. A range of lines and symbols referred to as hatchings are used to show different materials.

The main drawing that a bricklayer will use when setting out masonry for the substructure of a building is the floor plan. This view of the planned construction makes it relatively easy to interpret a range of information. It will show details such as internal wall positions, door and window positions, entry and exit of services, and other features that may have an impact on the way the substructure masonry is constructed.

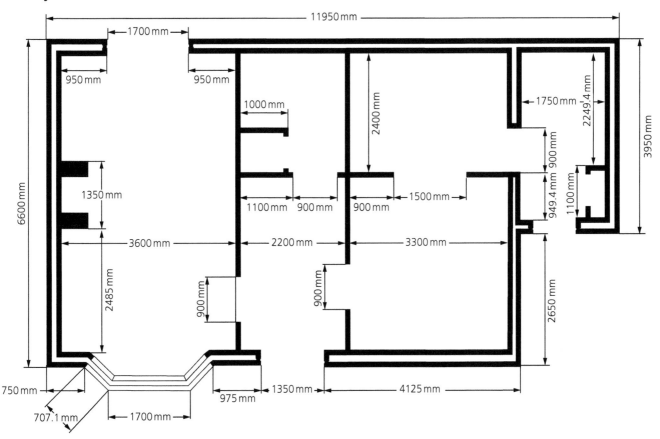

▲ Figure 2.7 Floor plan

To simplify the setting out of the masonry below ground, an additional view of the floor plan (sometimes referred to as a setting-out drawing) can be prepared. This provides a simple outline of the building's external walls with overall dimensions and measurements for positioning internal walls.

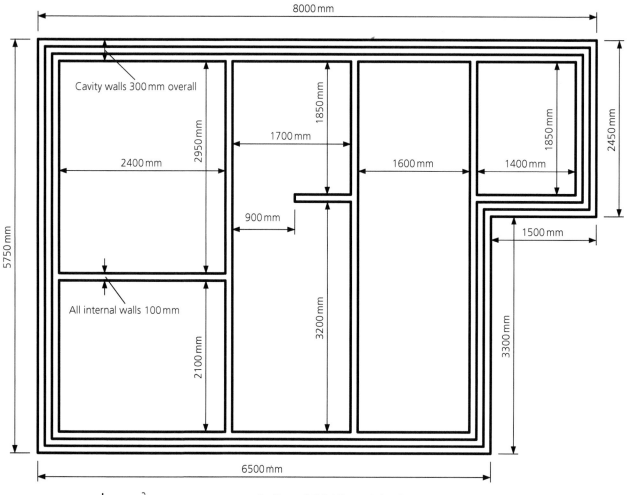

▲ Figure 2.8 Setting-out drawing

Although plans are drawn to scale, accurate setting out can only be achieved by working to the written dimensions, not by taking dimensions from the drawing using a scale rule. A scale rule has a range of markings that allow the user to read dimensions directly from a drawing that is produced to a stated scale. Only approximate dimensions should be 'scaled' from a drawing.

An example might be where drainage pipes for a toilet pass through a wall below ground. If the approximate position of the toilet is shown on the floor plan using a symbol, a dimension for positioning an opening for the toilet drain connection can be scaled from the drawing. An opening can then be created in a suitable position that is large enough to allow for later adjustments.

▲ Figure 2.9 A symbol shows the position of a drain for a toilet without a specific dimension

IMPROVE YOUR MATHS

If you do not have a scale rule, you can use an ordinary tape measure to establish approximate dimensions from drawings using the two most common scales of 1:50 and 1:100.

Carefully measure the required detail on the drawing in millimetres and multiply by 50 or 100 depending on the scale of the drawing. For example, if a line shown on a drawing measures 53 mm, for a scale of 1:50 multiply 53 by 50, which equals 2650 mm or 2.65 m.

When setting out a structure using dimensions from a working drawing, the bricklayer must be alert to the possibility of mistakes in measurements occurring during the production of the drawing. Make it a habit to check that individual dimensions for wall positions along the line of a wall add up to the overall dimension. This will quickly reveal any errors.

ACTIVITY

On a sheet of A4 paper, use a scale rule to draw a line to represent a length of 10 m for each of the scales in Tables 2.2 and 2.3. Is the paper big enough to draw a line at a scale of 1:5 or 1:10?

IMPROVE YOUR MATHS

Look carefully at the dimensions in Figure 2.10. Can you see a discrepancy between the individual dimensions and the overall dimension?

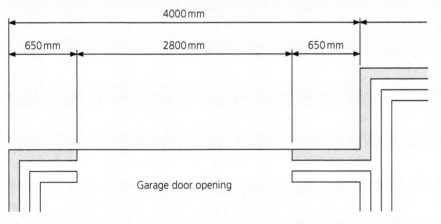

▲ Figure 2.10 Check that the individual dimensions add up to the overall dimension

If errors are identified, talk to your supervisor who has the responsibility to decide on the solution to the problem. On large projects, the supervisor or line manager may request written confirmation of the correct details from a more senior manager; on a smaller project, they may consult directly with the architect.

In addition to extracting information from working drawings, a bricklayer may obtain important details about a work task from other written documents.

Risk assessments and method statements are used to identify potential hazards and to provide information on safe and efficient working practices. For example, when setting out buildings, the ranging lines set out across the work area can create trip hazards, and when foundation trenches are dug, there is a risk of falling into the excavation. These are risks that must be considered.

INDUSTRY TIP

In Chapter 1, it is explained that drawings are described as 'contract documents', which means they are legally binding. That is why if you discover an error on a drawing you must consult with someone who has the authority to decide how to solve the problem.

Job sheets can be used to give specific information about a work task, such as the:

- location on site
- materials to be used
- number of personnel required to accomplish the task.

Information specific to setting out masonry structures can be accessed from a range of documents, including schedules and specifications. These documents are discussed in Chapters 3 to 6 in different contexts.

4 METHODS OF SETTING OUT AND TRANSFERRING LEVELS

Most buildings are set out as squares or rectangles. This means that positioning the profiles to set out a building will involve creating right angles (90°) at the corner positions.

4.1 Setting out right angles

There are several methods to accurately set out right angles, employing different tools and equipment.

The 3:4:5 method

Using a simple ratio, it is possible to set out 90° corners quickly and accurately using tape measures and string lines. By applying the ratio 3:4:5 to a right-angled triangle, we establish 90° angles for the corner positions of a building.

The ratio 3:4:5 refers to units of measurement. You can use any unit of measurement (metres, centimetres, millimetres) that is easy to work with and suits the needs of the job, as long as the ratio stays the same.

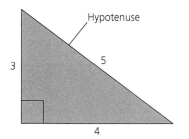

▲ Figure 2.11 Right-angled triangle

IMPROVE YOUR MATHS

Most smartphone calculators have a scientific calculator function. If you do not have a smartphone, borrow a scientific calculator, and try the following activity.

A small building is 2.5 m long and 1.5 m wide. Multiply 2.5 by 2.5 and write down the answer (this is called 'squaring' a number). Now multiply 1.5 by 1.5 and write down the answer. Add the two answers together.

Find the square root key on the scientific calculator. You might find it on an ordinary calculator; it will look something like this:

Enter the total of the two calculations and use the square root key. You should get the result 2.915. This is the measurement of the hypotenuse, the long side of a right-angled triangle, as shown in Figure 2.11.

Setting out a building using the 3:4:5 method is straightforward if simple steps are followed. Study the following step-by-step guide to see how this works.

Step by step
Using the 3:4:5 method to set out a right angle

Step 1 Position two pegs to represent the front of a building. Drive nails into their tops and attach a string line.

Step 2 Measure 3 m along the string line from the first peg and position a third peg with a nail at the exact dimension directly under the line.

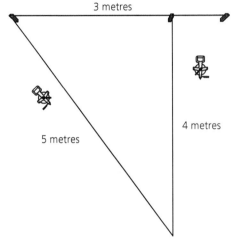

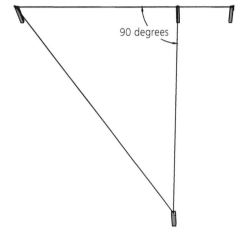

Step 3 Attach one tape measure (tape 1) to the nail on the first peg that was set up and attach the other tape measure (tape 2) to the nail on the third peg that was set up.

Step 4 Forming a triangle with the string line and the two tapes, read 4 m on tape 1 and 5 m on tape 2. Where these dimensions on the two tapes cross over each other, position a fourth peg, with a nail at the exact meeting point of the tapes.

Builder's square

A builder's square can be used as a guide to set up two string lines at 90° angles to each other. You can make a builder's square on site using available timber, creating a right angle using the 3:4:5 method described in the step-by-step guide.

> **INDUSTRY TIP**
>
> When setting out and positioning timber pegs, keep a carpenter's claw hammer in your tool kit. It will be easier to use than a club hammer to drive in nails. If you need to adjust the nail positions, you can use the claw hammer to pull out the nails to reposition them.

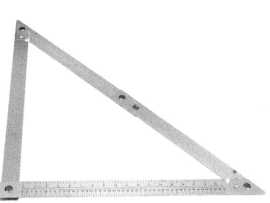

▲ Figure 2.12 Timber builder's square ▲ Figure 2.13 Metal folding builder's square

Try making your own miniature builder's square using strips of stiff card of a suitable width. The size is not important – just be careful to produce an accurate square, using the 3:4:5 method to create a right angle. Review the description of this method on the previous page to help you.

Alternatively, builder's squares are available that are manufactured from metal and foldable, which makes them more convenient to transport and store.

To produce accurate results when setting out walls at right angles using a builder's square, you must carefully align the string lines along the edges of the equipment. Slight inaccuracies in alignment will be magnified along the line of the walls, leading to construction of a building that is 'out of square'.

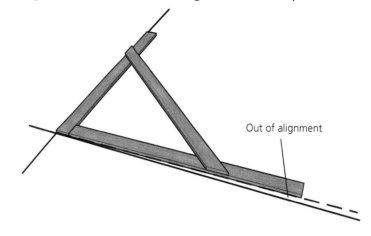

Out of alignment

▲ Figure 2.14 Careful alignment of string lines with a builder's square is vital

Optical square

An optical square is a simple instrument which has either:

- two sighting views set at right angles to each other, or
- an optical device known as a prism, which allows the user to view two points at right angles to each other at the same time.

The instrument is mounted on a tripod when in use, to keep it steady and accurately positioned plumb over a corner reference point. An assistant is directed by the person sighting through the instrument to mark two positions at right angles to each other.

This instrument makes setting out right angles a simple and accurate process.

▲ Figure 2.15 Optical square

4.2 Positioning profiles

Timber profiles are positioned at the corners of the building to be constructed. Most buildings will have loadbearing internal partition walls, which require a foundation to be set out for them. This means profile boards must also be provided at suitable intermediate points corresponding to the floorplan of the structure.

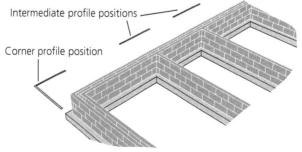

Intermediate profile positions

Corner profile position

▲ Figure 2.16 Intermediate profiles to set out internal walls

String lines are attached to the profiles to:

- guide the digging of trenches for the concrete foundation
- guide the bricklayer when positioning the walls of the building on the foundation concrete.

The front wall of the building (known as the **frontage line**) is set out first and must be located on or behind the **building line**. The side walls of the structure are set out at right angles to the frontage line, and the rear wall can then be set out parallel to the frontage line. These wall positions are established by positioning pegs at each corner.

KEY TERMS

Frontage line: the front wall of a building

Building line: a boundary line set by the local authority beyond which the front of a building must not project

INDUSTRY TIP

The frontage line is often confused with the building line. Keep in mind that the frontage line refers to the front wall of the building, which can be moved forwards or backwards. It could be positioned directly on, but never in front of, the building line. The building line cannot be moved – it is set by the local authority.

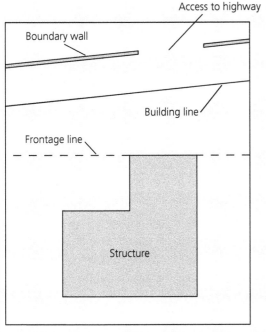

▲ Figure 2.17 Building line and frontage line

Profiles must be set out with care, to make sure that the finished building is positioned correctly on the site or building plot and that the outline and internal walls are located accurately.

The first step is to confirm the position of the building line by checking the block plan and site plan. Then two pegs can be positioned along the frontage line, corresponding to the corner positions of the building. Remember, the frontage line must never project in front of the building line.

Study the step-by-step guide to see the remaining stages of setting out the profiles ready for work to continue.

Step by step
Positioning profiles accurately

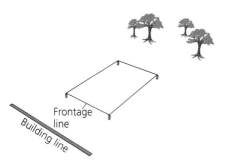

Step 1 Using one of the methods to create right angles previously discussed, set up pegs at the remaining two corners to create a rectangle.

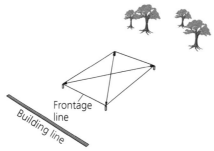

Step 2 Check the diagonal measurements are the same, to confirm the corners of the rectangle are 90°.

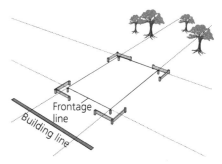

Step 3 Extend the lines of each side beyond the corner positions to set up corner profiles.

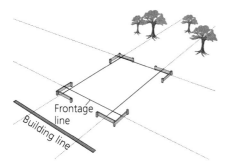

Step 4 Remove the corner pegs and attach lines to the wall positions on the profiles.

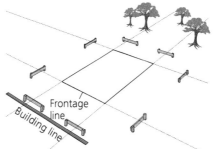

Step 4a Alternatively, if space is needed for a mechanical excavator to work, position profiles a short distance from the corner positions.

In Step 2, checking that the diagonal measurements are the same gives confirmation that the corners are set at 90° angles. However, this will only be the case if the overall length and width dimensions of the rectangular (or square) building are accurate. Check these for accuracy first.

ACTIVITY

If the overall length and width dimensions of a structure are measured accurately, but the diagonal measurements are not the same, the outline of a building will become what is known as a parallelogram.

Search online for images of parallelograms and then draw one yourself to get the idea of what a building that is out of square might look like.

4.3 Excavation of the foundation trench

Once the ranging lines have been attached to the profiles, markings can be made to guide the excavation of the foundation trenches. Guide lines are marked directly below the ranging lines for all walls that are to be excavated.

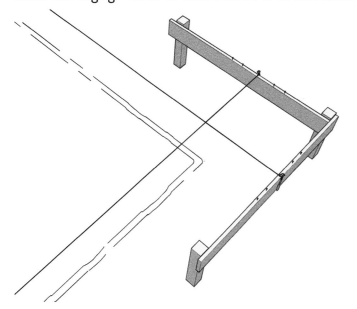

▲ Figure 2.18 Ranging lines in position with guide lines marked in spray paint

A commonly used method is to mark the centre line of the wall position, so that the excavator operator can easily align the centre of the digging bucket when excavating the foundation trench.

The top of the profile rail may have several nails showing a range of positions, including the centre line.

When the excavation of the trenches is completed neatly and the bottom of the trench has been cleaned out, the concrete foundation can be poured and levelled.

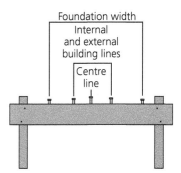

▲ Figure 2.19 Positions marked with nails on the top rail of a profile

4.4 Establishing wall positions

When the foundation concrete has hardened, the bricklayer can set out wall positions ready to construct the **footing** masonry. This is done by reattaching the ranging lines to the profiles to show the face lines of the building. The bricklayer then plumbs down from these lines to mark the wall positions onto the foundation concrete.

The traditional method of marking the wall position on the foundation is to spread a thin layer of mortar (called a screed) on the surface of the concrete directly underneath the ranging lines. This can be marked with the tip of a trowel to follow the face line of the wall. Some bricklayers use a line of spray paint to provide a clean surface for marking with a permanent marker pen.

KEY TERM

Footing: the section of masonry from the concrete foundation to the ground-floor level; sometimes the whole foundation is referred to as 'footings'

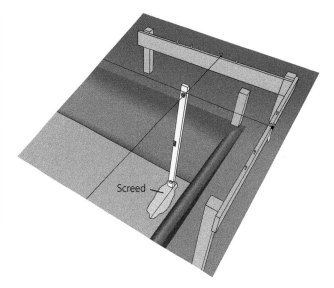

▲ Figure 2.20 Plumbing down from ranging lines to the foundation concrete

When the lines of wall positions at the corners of the building have been carefully marked on the foundation concrete, the bricklayer has a reference to build to which matches the string lines attached to the profiles. Be careful to lay the first bricks or blocks on the correct side of the marked lines. Develop the habit of continuously checking your work for accuracy.

4.5 Establishing the correct level of a building

To establish the correct level of a building, a reference point called a datum is set up on site. This is often referred to as a temporary bench mark (TBM). The TBM can take the form of a peg which is secured in concrete and protected from any disturbance that could be caused by vehicle or machinery movements or other construction activities.

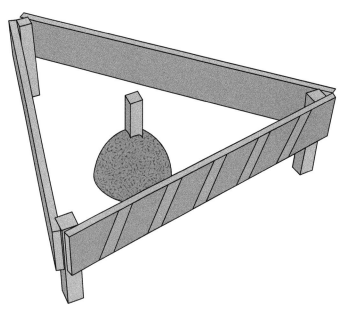

▲ Figure 2.21 Datum point protected from disturbance

It is important that the datum is not disturbed, since all levels for buildings are transferred from it. Sometimes a fixed point, such as a marked point on a roadside kerb-stone or an inspection chamber cover in a road near to the site, may be used as a datum.

Usually, a level is transferred from the datum to the location of the building being constructed to establish the finished floor level (FFL) as the first reference point to build to. This usually corresponds to the damp proof course (DPC) level. Chapters 1 and 5 give more detail on these elements of the building.

4.6 Levelling equipment

A range of equipment can be used to transfer levels. Some equipment is quite basic, while other tools are technically advanced. Whichever type of equipment is used, care is needed to avoid errors and discrepancies in establishing levels.

Spirit level and straight edge

A traditional method of transferring levels makes use of relatively simple equipment; a spirit level and straight edge can be used together to transfer levels from one point to another.

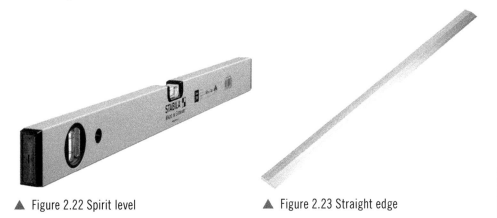

▲ Figure 2.22 Spirit level   ▲ Figure 2.23 Straight edge

When transferring levels over distance on site, this method requires care and can be time consuming. Levelling over a long distance requires a series of pegs, spaced at a measurement that is slightly shorter than the length of the straight edge being used. The method is best suited to transferring levels over shorter distances.

When using a spirit level and straight edge together, it is important to 'reverse' them end to end between levelling pegs. This will largely cancel out any small inaccuracies in the spirit level or distortions in the straight edge. This is especially important if the straight edge is made from timber, which can warp or twist over time. Study the following step-by-step guide to see how this is done.

> **INDUSTRY TIP**
>
> Straight edges manufactured from a lightweight metal such as aluminium are much more stable than timber straight edges. Timber is affected by changes in temperature and moisture levels, so a timber straight edge must be checked frequently to make sure it is not changing shape.

Step by step
Transferring levels using a spirit level and straight edge

Step 1 Carefully level between the first two pegs, making adjustments to the second peg in the row.

Step 2 Reverse the spirit level and straight edge end to end and carefully level between the second and third pegs. Only make adjustments to the third peg.

Step 3 Continue reversing the spirit level and straight edge between subsequent pegs.

Optical level

An optical level is a more technically advanced piece of equipment that is efficient and accurate for transferring levels over long distances on site. Care is needed when setting up to ensure the head of the instrument is perfectly level and the tripod is stable.

Two operatives are required to transfer levels – one to 'sight' the optical level and one to position a staff with measurement markings on it (a graduated staff), from which the level readings are taken.

Always inspect the instrument visually before use to make sure there is no evidence of damage. It must be handled and stored carefully and protected against dust and vibration to maintain its accuracy.

▲ Figure 2.24 An optical level must be set up carefully

▲ Figure 2.25 One operative takes a height reading with the optical level

▲ Figure 2.26 Another operative positions the graduated staff

Laser level

Accurate transfer of levels can be achieved easily with an electronic instrument such as a laser level. This instrument is often self-levelling and can be used by one operative working alone. Once the instrument has been set up on its tripod, the operator is free to move anywhere on site carrying a staff-mounted receiver or detector.

▲ Figure 2.27 Laser level

Using a laser level, the reference level from the TBM can be easily and accurately transferred to multiple positions over great distance. Always be aware of the potential hazard to eyesight posed by laser light when operating this type of equipment.

4.7 Setting the level of the masonry

The specified level is transferred to datum pegs at each corner of the building and to any other points where a reference for level is required. The datum pegs can be positioned at the same location as the corner profiles. If the corner profiles are at a suitable height, a nail can be positioned on them to serve as a datum.

The bricklayer can then use a spirit level and tape measure to **gauge** down from the datum to the top of the foundation concrete. This allows the bricklayer to work out how many **courses** of bricks or blocks will be needed to build the footing masonry to the FFL (or DPC) level.

ACTIVITY

Look for laser levels online. Note the name and price of the cheapest and the most expensive laser level you can find. You may be surprised at the range of costs.

KEY TERMS

Gauge: in this context, the process of establishing measured uniform spacing between brick or block courses including the horizontal mortar joints

Courses: continuous rows or layers of bricks or blocks on top of one another

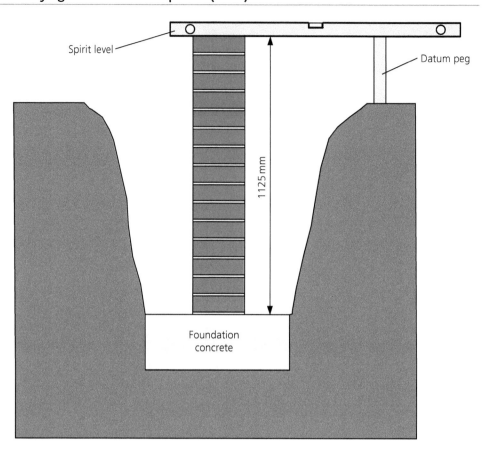

Spirit level

Datum peg

1125 mm

Foundation concrete

▲ Figure 2.28 Gauging down from a datum peg to the top of the foundation concrete

INDUSTRY TIP

When measuring down to the foundation concrete from a datum peg, it can be easier to use a gauge rod instead of a tape measure. This is a timber rod used to check that brickwork and blockwork are constructed to accurate height measurements with consistent joint sizes. It is sometimes referred to as a 'lath' rod.

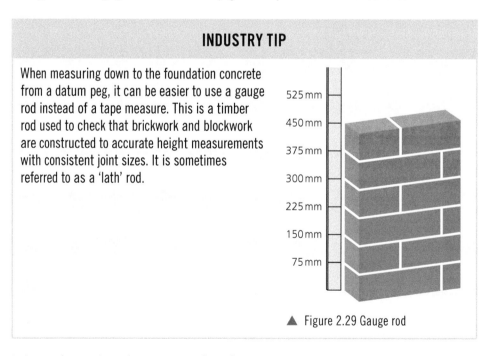

525 mm
450 mm
375 mm
300 mm
225 mm
150 mm
75 mm

▲ Figure 2.29 Gauge rod

It is good practice when pouring foundation concrete to set the level at a height that allows for full courses of brick or block between the top of the concrete and the datum. If the level is not set to accommodate gauged masonry, a lot of cutting of bricks and blocks will be required, which will waste both time and resources.

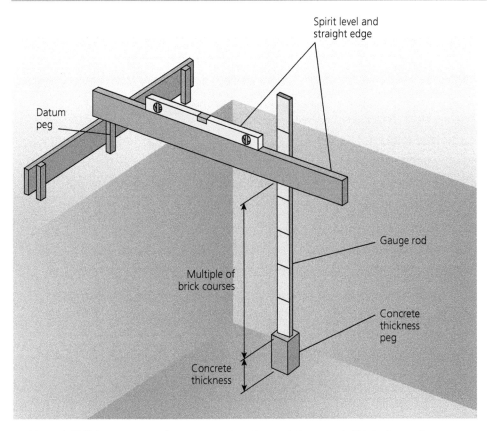

▲ Figure 2.30 Using a gauge rod to measure down from a datum peg; in this instance, the thickness of the foundation concrete is being set from the datum

Always keep in mind that the process of setting out a building must be done with care. Accurate measuring and levelling from given reference points is vital to ensure the building is set out in accordance with the specified design and positioned correctly in relation to its surroundings.

Test your knowledge

1 Which type of drawing shows the proposed development in relation to the site boundary?

   a Block plan

   b Detail drawing

   c Site plan

   d Section drawing

2 Which complete section of a building is called the substructure?

   a Below ground-floor level

   b Below first-floor level

   c Above foundation level

   d Above ground-floor level

3 What are the temporary timber guides called that are used for setting out a building?

   a Positions

   b Profiles

   c Pointers

   d Patterns

4 Which letters are used to indicate a central datum point on site?

   a TBN

   b TMB

   c TBM

   d TNB

5 Which angle is set out using an optical square?

   a 30°

   b 45°

   c 60°

   d 90°

6 To which scale is a detail drawing produced?

   a 1:5 or 1:10

   b 1:50 or 1:100

   c 1:100 or 1:200

   d 1:1250 or 1:2500

7 What is the boundary line set by the local authority beyond which a building must not project when setting out?

   a Frontage line

   b Ranging line

   c Building line

   d Guide line

8   Which instrument can be used by a lone operative to transfer levels
    on site?

    a   Optical square

    b   Laser level

    c   Optical level

    d   Laser pointer

9   What is the method called when a thin layer of mortar is used to mark
    wall positions on foundation concrete?

    a   Scratch

    b   Scrape

    c   Screed

    d   Score

10 Which side of a right-angled triangle is called the hypotenuse?

    a   The shortest side

    b   The left side

    c   The right side

    d   The longest side

CHAPTER 3

CARRYING OUT BLOCKLAYING ACTIVITIES

INTRODUCTION

Many masonry structures are constructed in both brick and block, but there are occasions when only blocks are specified as the materials for a work task. This chapter will focus on the skills and techniques that must be developed to use blocks safely and efficiently as a construction material.

To build successfully, it is important to understand the characteristics of different types of block used in masonry construction. Selection of the correct tools and equipment will also be discussed, along with methods of accurately setting out and building straight walls, corners and junctions using blocks.

LEARNING OUTCOMES

After reading this chapter, you should:
1. know how to prepare for blocklaying activities
2. know how to identify and select the correct resources for blocklaying activities
3. know how to build straight walls, corners and junctions using blocks.

1 PREPARATION FOR BLOCKLAYING ACTIVITIES

With many activities in life, the key to success is good preparation. This is the case when building a wall using blocks. A successful bricklayer must develop skills in planning and organisation, as well as learning to work safely and efficiently with others.

Safety and efficiency are closely linked. Two documents that should be consulted when preparing a work task are:

- risk assessments
- method statements.

Risk assessments safeguard workers from potential hazards and method statements advise on safe and efficient sequences of work, allowing workers to plan and prepare confidently.

▲ Figure 3.1 Check risk assessments and method statements when preparing for work

1.1 Common hazards during preparation

Good preparation includes getting into the right mindset about health and safety. All personnel on a construction site are responsible for making sure they work in a safe manner, to protect themselves and those around them. For example, you should always use the correct personal protective equipment (PPE) when undertaking any construction tasks. (Chapter 6 discusses PPE in more detail.)

There are many potential hazards on a construction site that can be managed or eliminated by being safety conscious during preparation and planning.

Hazardous materials

Some materials used in construction must be treated with care to avoid injury. For example, using wet concrete blocks can cause irritation to a bricklayer's skin due to the release of lime from the cement. Keeping the blocks dry and wearing suitable gloves will protect the bricklayer.

Contact with cement in a mortar mix for long periods can cause skin conditions such as dermatitis or even chemical burns.

The Control of Substances Hazardous to Health (COSHH) Regulations 2002 must be followed when using potentially hazardous materials and are an important source of information to protect workers from injury. Manufacturer's instructions and safety policy documents should also be viewed when preparing for a work task, since they can provide safety information on specific materials.

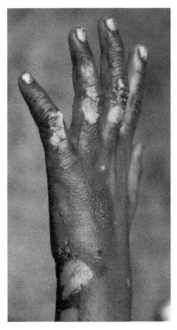

▲ Figure 3.2 A chemical burn as a result of not wearing safety gloves

Moving and stacking materials

Blocks are often heavy and they can be awkward to handle. When moving and stacking blocks, potential hazards include crushing injuries, muscle strains, and cuts and bruises to exposed areas of skin.

▲ Figure 3.3 Be careful when stacking blocks that you do not trap your fingers

INDUSTRY TIP

The risk of injury is largely determined by the weight of the block – the heavier the block, the higher the risk of injury. Look for ways to minimise manual handling and plan to use mechanical handling provisions on site as much as possible, for example, forklifts and cranes.

Keep the following points in mind to minimise the risk of injury, especially when moving and handling blocks manually:

- Operate a two-person system for blocks weighing more than 20 kg.
- Arrange the work activity to avoid overreaching or twisting the body.
- Ensure a good grip on the blocks and a secure foot placement in the working area.
- Arrange the work task so that blocks are only handled up to shoulder height.

Mixing mortar

A drum mixer will often be used on site to mix large quantities of mortar quickly and efficiently. Various sizes of mechanical mixer are available to suit the production needs of the job. These are powered either by electricity or by a diesel/petrol engine.

Any size of drum mixer is a powerful machine and must be treated with respect. Be aware of potential hazards such as loose clothing getting trapped in moving parts – carelessness can lead to severe injury.

Later in this chapter we will discuss methods of mixing mortar by machine and by hand in more detail.

▲ Figure 3.4 A general operative loading a mixer

Cutting blocks

Construction tasks of any size usually require blocks to be cut and trimmed. Cutting blocks by hand can be hazardous due to the creation of dust particles and flying block fragments that can cause eye injuries. This can be a risk to both the operative carrying out the work and those working nearby, so develop the habit of considering safety requirements before starting any cutting operations.

Take care when using hammers and steel chisels to cut blocks. The strong hammer blows required during cutting operations can cause hand injuries.

ACTIVITY

Search online for 'faults in construction hand tools'. Make a list of common faults that can develop in hammers and steel chisels.

Insulation blocks are manufactured from a lightweight material that can be cut using either a hammer and bolster or a masonry hand saw. Cutting these types of block can create a lot of fine dust, which is a potential health hazard. Not only can fine particles cause irritation to the eyes, nose and throat, but they can also cause damage to the lungs.

Methods of cutting blocks safely and accurately are discussed in more detail later in this chapter.

▲ Figure 3.5 Always use the correct PPE

KEY TERM

Induction: an occasion when someone is introduced to and informed about a new job or organisation

INDUSTRY TIP

When you start work on a new site, you will take part in an **induction**, which will alert you to specific hazards you need to be aware of. Take note of whom you should speak to if you have concerns about health and safety issues.

1.2 Documents used when preparing

Important and detailed information about the work task is provided by the specification, for example:

- the type and quality of materials to be used
- working practices that must be employed for a specific job to achieve the required quality of finish.

As mentioned in Chapter 1, where there are repeating components or features in a work task (i.e. they are used for several jobs or locations on site), they may be listed in a document known as a schedule. This document clearly links the listed components and features to specific work tasks in different buildings or site locations. See Figure 1.12 on page 8 for an example of a schedule.

ACTIVITY

The full range of documents used in construction activities is discussed in Chapter 1. Review the chapter and list all the documents you can find, briefly describing each one.

Using information sources correctly when preparing for the work task is vital to avoid mistakes that will be expensive to put right later. Building regulations will be applied when designing a structure, and these regulations will be reflected in the information documents a bricklayer must work to.

1.3 Drawings used when preparing

A bricklayer will also need to refer to working drawings when preparing for blocklaying tasks. These drawings can communicate a lot of information, avoiding the confusion that could potentially be created by lengthy written instructions.

▲ Figure 3.6 A working drawing can provide a great deal of information

Working drawings usually show a scale representation of the work task. Drawing to scale means that the work task is represented with accurate proportions but reduced in size to fit onto a manageable size of paper. If a full-size drawing of a project such as a house was produced, it would be almost impossible to use on site.

Scale is shown using a ratio, such as 1 to 10 (usually written as 1:10). On a drawing with a scale of 1:10, a wall that is 1 m (1000 mm) long in real life will be represented by a line that is 100 mm long. This is because 100 mm is one tenth of 1000 mm.

IMPROVE YOUR MATHS

A rectangular building is to be constructed that is 2.5 m long and 1.5 m wide. On A4 paper (in landscape), use a ruler to draw this building to a scale of 1:10.

Chapter 1 describes a complete range of drawings used in construction activities. The main types of drawing used by a bricklayer for blockwork tasks are listed in Table 3.1.

▼ Table 3.1 Types of drawing used for blockwork tasks

Type of drawing	Description	Example
Plan	A view from above the work task (think of it as a bird's-eye view) Scale 1:50 or 1:100	Plan view of a single-leaf block wall 
Elevation	A view showing each face or side of a work task Scale 1:50 or 1:100	Front elevation of a single-leaf block wall 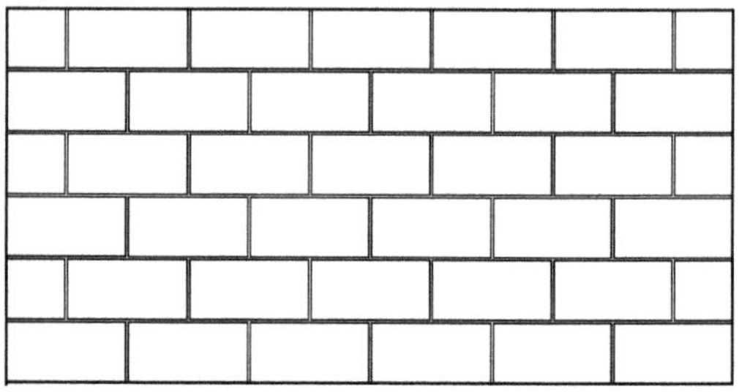
Section	A slice through a work task, which can show details that would otherwise be hidden Scale 1:50 or 1:100	Section view of a single-leaf block wall 

Note: specific details can be shown at a larger scale of 1:10 in any of the above drawing types to make things clearer.

INDUSTRY TIP

Notice the criss-cross hatchings in the section drawing. These show that the material within the 'slice' of the wall is block.

Some features of a job may not be shown on a drawing using specific dimensions. An example would be the entry point in blockwork footings for services such as water, gas, electricity or drainage. A bricklayer could use a scale rule to establish approximately where to leave an opening big enough to allow for adjustments to the position of the services during installation.

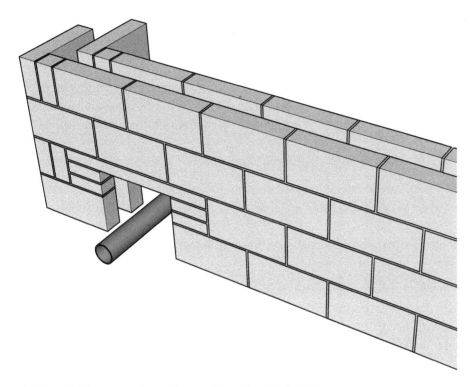

▲ Figure 3.7 An opening for services positioned in a block footing

However, where specific dimensions are given, you must always follow them when setting out measurements. Attempting to establish accurate measurements from a drawing by using a scale rule can lead to inaccuracies.

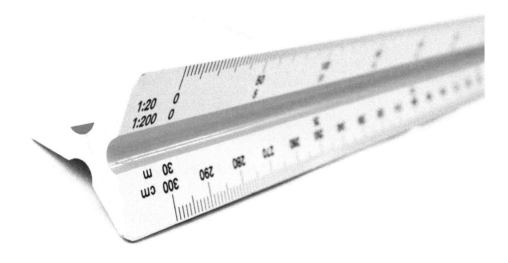

▲ Figure 3.8 Scale rule

2 IDENTIFYING AND SELECTING THE CORRECT RESOURCES FOR BLOCKLAYING ACTIVITIES

It is good practice to make lists of the materials required for a job. Consulting the specification and schedules will help you to identify the correct resources and calculate required quantities.

Making sure the correct quantities of materials are available before starting work reduces the possibility of running out of resources mid-project. Time will be wasted if the job must be interrupted to bring additional materials to the work location.

2.1 Calculating quantities

Chapter 1 discussed in detail the ways in which calculations of volume and area are used in construction. In this chapter on blocklaying activities, calculations of area are used. This is the first calculation when working out the quantity of blocks needed for a work task. To calculate the area of the face of a wall, you simply need to multiply its length by its height.

Although measurements on a working drawing are normally stated in millimetres, calculation of the surface area of a wall is done in m² (square metres). Since each m² of a single-leaf block wall contains 10 blocks, to calculate the number of blocks required to build a wall, you simply multiply the number of m² by 10.

IMPROVE YOUR MATHS

Work with a partner to create dimensions for imaginary block walls. Use realistic dimensions to a maximum of 5.5 m long and 2.5 m high and calculate the area of the wall face. Use the area figure to calculate the number of blocks required.

Using the length of a wall as one of the dimensions when calculating area is an example of using linear measurement. This type of measurement is used when an overall dimension is split up into smaller parts, for example, where an opening is left in a length of wall to allow for a gate. Study Figure 3.9 to see how this works.

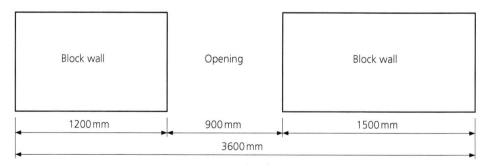

▲ Figure 3.9 Linear measurements along the length of a wall

When calculating the quantity of blocks required, it is important to add a **percentage** for waste. Breakages can occur when moving and handling blocks to transport them to the work location. The broken pieces that are not usable will have to be disposed of, resulting in waste. Also, when blocks are cut, the offcuts may not be usable and will become waste.

As a rule of thumb, add 5% to the quantities calculated to allow for wastage.

IMPROVE YOUR MATHS

To calculate 5% for waste is simple. Take the calculated number of blocks, multiply by 5 and divide by 100 (remember 'per cent' is an expression of hundredths).

So, if 300 blocks are needed for a work task, the calculation would be:

300 × 5 = 1500

1500 ÷ 100 = 15

5% for wastage is 15 blocks. The total number of blocks required, allowing for wastage, is 315.

ACTIVITY

Find an online calculator for working out quantities of mortar (search for 'mortar calculator') and try it out. Measure a local wall and then use the dimensions to work out the required quantity of mortar.

2.2 Checking suitability

Carefully checking the condition and suitability of materials and resources before starting construction activities maintains high standards and supports good levels of productivity. For example, if damaged blocks are used, the appearance and strength of a wall could be affected.

▲ Figure 3.10 Check the suitability of materials before you use them; these lightweight thermal blocks have been damaged by careless handling

Blocks are usually delivered to site in packs that are securely held together by steel or heavy-duty plastic bands. This makes them easier to move to the work location using a forklift or crane and reduces the likelihood of damage occurring during transport.

Materials should be protected from bad weather, both when in storage on site and when transported to the work location. If blocks are laid when they are wet, chemicals can be released which can cause long-term damage to the masonry and frost damage can occur in freezing weather.

2.3 Range of materials

Blocks

There are several types of block used in building walls of various thicknesses at different locations in a structure. Blocks are manufactured to standard dimensions of 440 mm long by 215 mm high. The width (or thickness) ranges from 75 mm to 215 mm.

Dense concrete blocks can be used to build walls that form part of a foundation or footing below ground. They can also be used above ground in single-leaf walls or cavity walls (see Chapter 5 for more on cavity walls). They are manufactured in concrete, which makes them suitable for load-bearing locations in a building.

Partition walls that form rooms in a building and party walls between properties can be built in dense concrete block. Because of the nature of the material they are manufactured from, dense concrete blocks have good fire resistance and sound insulation properties.

▲ Figure 3.11 Dense concrete block

Lightweight concrete blocks are manufactured from materials that make them easier to handle. They can be specified to carry certain loadings in given locations in a building design.

Lightweight insulation blocks are made from materials that are **aerated**. Tiny air bubbles of a controlled size are formed in the materials during manufacture. These bubbles then act as an insulator in the finished product.

▲ Figure 3.12 Lightweight insulation blocks

IMPROVE YOUR ENGLISH

Research insulation blocks online to find out about the manufacturing process and the materials used. Write down why the manufacturing process makes them good heat insulators.

Some blocks made from this aerated material are suitable for use below ground, but in the majority of cases they are used in the **superstructure** of a building.

There is more about the characteristics of blocks in Chapter 5, where their use in cavity walls is discussed.

Mortar

Mortar is the material used to form the horizontal and vertical joints that bond blocks together. It is mainly composed of **well-graded sand** and ordinary Portland cement (often referred to as 'OPC') mixed to a specified ratio. There are other types of cement which you will learn about at Level 2.

KEY TERM

Well-graded sand: sand that has large, medium and small grains

KEY TERMS

Aerated: exposed to the circulation of air

Superstructure: the section of a building from the ground-floor level upwards

Large grains:
40% (2.5–5 mm)

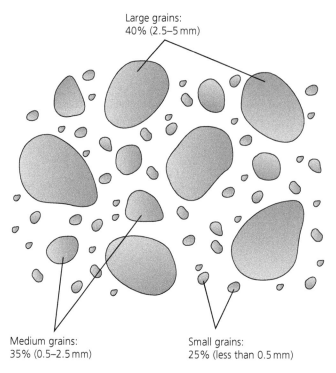

Medium grains:
35% (0.5–2.5 mm)

Small grains:
25% (less than 0.5 mm)

▲ Figure 3.13 Graded grains of sand

The specified ratio will vary depending on the type of blocks being used. A common mortar ratio is 1:4, which means that one part cement will be mixed with four parts sand. Increasing or decreasing the proportion of sand will make a weaker or stronger mix. Dense blocks need a stronger mix ratio, whereas lightweight insulation blocks need a weaker mix ratio.

Simply mixing sand and cement with water will produce a mix that is difficult to use, so a plasticiser is added to make the mix more pliable and easier to work with. This is usually in the form of a liquid or powdered chemical additive that traps tiny, evenly sized bubbles of air in the mix (called air entrainment), allowing the grains of sand to move over each other more freely.

▲ Figure 3.14 Liquid plasticiser

▲ Figure 3.15 Powdered plasticiser

Traditionally, powdered lime was used in mortar, which acted as a plasticiser before it hardened. This is often referred to as 'hydrated' lime because of the manufacturing process. Increasing the powder content in the mortar forms a paste that allows sand grains to move more freely over each other to increase workability. However, there are safety hazards to consider when using lime, as it can result in chemical burns. It tends to be used less often since the introduction of safer, easier-to-use chemical additives.

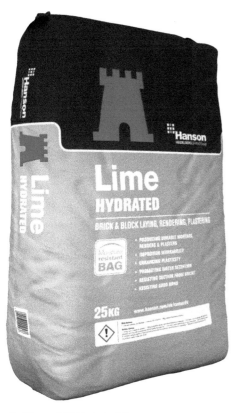

▲ Figure 3.16 Hydrated lime

> **HEALTH AND SAFETY**
> Search online for 'safe use of hydrated lime'. Write down the potential hazards of using lime powders on site.

INDUSTRY TIP

Some bricklayers use washing-up liquid as a plasticiser. This is not good practice, as it produces bubbles that are not controlled in size and which form tiny pockets in the hardened mortar. Moisture can be retained in these tiny pockets; if this freezes, it can severely damage the masonry.

2.4 Range of tools

Before starting work, it is good practice to list the tools and equipment needed for a masonry task. This contributes to efficiency.

Tools can be split into three main categories:

- laying and finishing
- checking
- cutting.

The tools from these three categories needed for blocklaying are listed in Table 3.2.

▼ Table 3.2 Tools for blocklaying

Category	Tool	Description
Laying and finishing	Trowel	Used for forming mortar joints when laying blocks
	Pointing trowel	Used for filling gaps in mortar joints Can be used for smoothing mortar applied to the top of sections of masonry
	Jointer	Used for forming a half-round or concave joint finish to mortar joints Sometimes referred to as a jointing iron
Checking	Tape measure	Used to set out and check dimensions
	Spirit level	Used to check work is plumb and level

▼ Table 3.2 Tools for blocklaying (continued)

Category	Tool	Description
Checking	Line and pins	Used as a guide to accurately lay blocks along the length of a straight wall
	Corner block	Used to position a line and pins on corners (or quoins) – the tension of the string line stretched from end to end keeps the corner blocks in position
Cutting	Club hammer	Used together with a brick bolster to accurately cut blocks to the required dimensions Sometimes called a lump hammer
	Bolster	Used like a chisel with a club hammer to cut blocks accurately to the required dimensions

▼ Table 3.2 Tools for blocklaying (continued)

Category	Tool	Description
Cutting	Brick hammer	Used to cut and trim blocks quickly where accuracy is not critical
	Scutch hammer	Used to accurately and finely trim and shape blocks as required (more frequently used to trim and shape bricks) Sometimes referred to as a comb hammer

2.5 Range of equipment

Table 3.3 shows some of the equipment used for blocklaying.

▼ Table 3.3 Equipment used for blocklaying

HEALTH AND SAFETY

Check the condition of your tools and equipment before use. Using tools that are not in a safe condition can cause injury to yourself and others nearby.

Maintaining your tools carefully not only keeps you and those working around you safe but it will also make the tools last longer.

Equipment	Description
Spot boards	Boards that have mortar placed on them within easy reach of the bricklayer along the length of a wall
Shovel	Used to move mortar and loose materials or for mixing mortar by hand

▼ Table 3.3 Equipment used for blocklaying

Equipment	Description
Wheelbarrow	Essential for moving materials to the work location
Gauge rod	Used to check that blockwork is constructed to accurate height measurements and with consistent joint sizes Sometimes referred to as a lath rod
Builder's square	Used for accurately setting out work with right-angled corners
Straight edge	Used to transfer levels over distances longer than a spirit level Can be made from timber but must be regularly checked for accuracy because timber can change shape in varying weather conditions; a more stable material is a lightweight metal such as aluminium, as shown in the illustration
Tingle plate	Used to support a string line stretched over long distances (a long string line could sag in the middle; if a wall is built following the line, it will also sag in the middle)

Tingle plate

INDUSTRY TIP

Sweeping brushes, hand brushes and buckets are also useful items of equipment for blocklaying tasks.

3 BUILDING STRAIGHT WALLS, CORNERS AND JUNCTIONS IN BLOCK

Chapter 5 covers in detail important points regarding positioning materials and components ready for work. Consideration is given to methods of safely and efficiently moving and stacking blocks ready for building.

▲ Figure 3.17 Stacks of blocks and mortar spot boards set out ready for work

In this chapter, we will concentrate on how to set out blockwork to form straight single-leaf walls, corners and junctions. The first task in the building sequence is to prepare and mix mortar to the specified mix ratio and consistency.

3.1 Mixing mortar

As mentioned previously, mortar can be mixed as required on site using a drum mixer. Water is added to the drum first, followed by the dry materials, in accordance with the specified ratio. Adding water first avoids the dry materials sticking to the sides of the drum, which would prevent thorough mixing of the mortar. The consistency is adjusted by the careful addition of small amounts of water as required.

A more modern approach to mixing on site is to use a **dry silo mixer**. The silo mixer has the advantage of producing mortar of dependable quality and consistency, almost at the push of a button.

KEY TERM

Dry silo mixer: a machine that contains all the dry materials to produce precise quantities of mixed mortar on demand

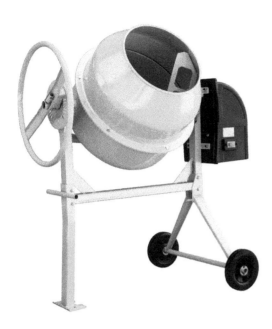

▲ Figure 3.18 Drum mixer

▲ Figure 3.19 Dry silo mixers

Using a drum mixer or silo mixer minimises waste, since they both produce mortar as it is needed, to be used straight away. An alternative is to have mortar delivered to site which has been pre-mixed at an off-site plant. This method of production requires a chemical additive called a retarder to slow down the setting time, so that the material remains workable throughout the working day.

IMPROVE YOUR ENGLISH

Think about the different ways of mixing mortar: drum mixing, dry silo mixing, pre-mixing. Write a list of advantages and disadvantages for each method.

There will also be occasions when mortar needs to be mixed by hand, for example, for smaller jobs or where a work task is difficult to access with mixing machinery. Mixing mortar by hand is hard work, but it can be made easier by following a simple process.

A sequence often used when mixing by hand is known as 'three times dry, three times wet':

- Move the dry materials from their placed position to the side and then back again three times to mix the sand and cement. It is best to work on a smooth, level base if possible.
- Once the dry materials have been mixed, form a large circular dip in the centre of the mound and add water carefully. Gradually mix in the water, until it is evenly spread through the materials.
- Turn the mixture three times 'wet', in a similar manner to the dry mixing process, and add more water slowly as required. If a plasticiser is used, it should be added to the water and not to the dry materials.

INDUSTRY TIP

The quality of the water used to mix mortar is very important. It must be pure enough to drink (referred to as 'potable' water). Water of this quality will be free from chemicals or other substances which could affect the hardening process of the mortar.

▲ Figure 3.20 Water for mixing mortar must be clean enough to drink

Study the step-by-step guide to see this method of mixing mortar by hand.

Step by step
Mixing mortar by hand

Step 1 Add cement to the sand in the proportions specified by the ratio.

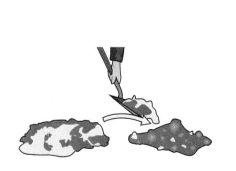

Step 2 Mix the sand and cement together by moving them to the side and back three times.

Step 3 Add water to a centrally formed dip in the dry mixed materials, taking care not to add too much.

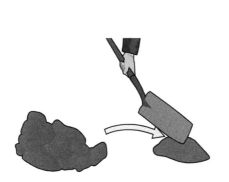

Step 4 Make sure the water is distributed evenly through the mound of materials.

Step 5 Mix in the water completely by moving the pile to the side and back three times.

Step 6 Add small amounts of water as needed to improve the consistency.

Keep in mind that the mix ratio and consistency of the mortar will vary, depending on the type of block being used. For example, a dense concrete block is less absorbent than a lightweight insulation block and will react to temperature changes differently, so the mortar will be mixed to a different sand/cement ratio and consistency.

3.2 Setting out a straight block wall

Having carried out careful preparation work, you must maintain high standards in the actual construction of the wall. Develop the habit of continuously checking your work to ensure quality and to become more proficient and skilled.

A straight, single-leaf wall using blocks will commonly be half-bonded. This means that each alternate course is laid with an overlap of half the length of a standard block, so that the **perp** joints in any one course are exactly halfway along the face of the blocks in the course below.

KEY TERM

Perp: short for 'perpendicular'; the vertical mortar joint at right angles to the horizontal mortar bed joint

▲ Figure 3.21 A half-bonded block wall with plumb perps

▲ Figure 3.22 A poorly bonded block wall

To make sure the perp joints are the correct size over the length of the wall, it can be useful to set out the wall 'dry', which means placing the blocks in position without any mortar. You can then check the joint sizes and either tighten or open them slightly so that the wall will fit within the overall dimensions.

▲ Figure 3.23 Setting out the bond 'dry'

If the resulting joint sizes are too big or too tight and the wall does not fit within the specified dimensions, the bricklayer will have to use other methods.

One method is to use **reverse bond**. If there are quoins (corners) at each end of the wall, rather than expecting the wall to have blocks in the same direction at either end, the bonding arrangement is reversed, so that at one end a block will be in line with the wall and at the other end a block will be at right angles to the wall.

If the wall has **stopped ends**, reversing the bond will mean that one end has a full block while the other end has a half block.

KEY TERMS

Reverse bond: for a wall with corners at each end, one end will have a block in line with the face of the wall and the other end will have a block at right angles (90°) to the face of the wall; for a wall with stopped ends, one end will have a full block and the other end will have a half block

Stopped end: squared-off vertical finish to the end of a wall

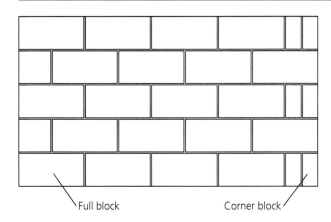

▲ Figure 3.24 Reverse bond in a block wall with a corner

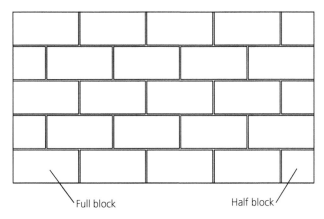

▲ Figure 3.25 Reverse bond in a stopped-end block wall

If reversing the bond does not produce a wall with uniform joints within the overall length measurement, then the only choice is to introduce cut blocks in each course. Placing cut blocks in a course is called **broken bond**.

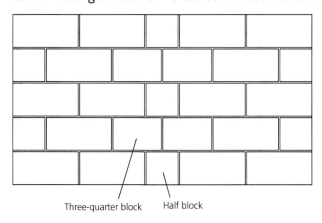

▲ Figure 3.26 Broken bond in blockwork

Cut blocks in broken bond should be placed as near to the centre of the wall as possible. When a half block is used in a course, the courses above and below it will contain two three-quarter cuts to maintain the bond.

KEY TERM

Broken bond: the use of cut blocks (or bricks) to establish a good bonding pattern where full blocks (or bricks) will not fit

INDUSTRY TIP

A block wall may have a lot of cuts of the same size to maintain the bond. It speeds up the job if you cut all the blocks needed in one operation before you start laying.

3.3 Cutting blocks

Before cutting blocks, consider the potential hazards. Make sure nearby workers are also aware of the risks. As well as using basic PPE, such as a hard hat and safety boots, wear safety glasses or goggles to protect your eyes from flying block fragments.

In addition, when machines are used for cutting, wear a dust mask as breathing protection and ear defenders or ear plugs as hearing protection.

▲ Figure 3.27 Goggles

▲ Figure 3.28 Safety glasses

▲ Figure 3.29 Ear defenders

▲ Figure 3.30 Ear plugs

When cutting dense or lightweight concrete blocks by hand, a club (or lump) hammer and bolster chisel are used to quickly produce a neat cut. A pencil and tape measure are used to accurately mark the position of the cut.

The blade of the bolster chisel can then be carefully positioned slightly on the waste side of the pencil line and the chisel firmly struck with the hammer. If adjustment of the cut edges is needed, a brick hammer can be used to trim the block to shape.

Sometimes blocks are used to form the face of a wall, so the blockwork will be permanently visible. This means that the standard of finish must be higher than if the blockwork is hidden. Cutting operations therefore need to be more precise and a scutch hammer may be used to accurately trim cut blocks to shape. An example of this might be if the blocks in a face wall are cut to follow the angle of a pitched roof.

As you travel around, look for examples of blocks used in face walls. A sports stadium or leisure centre will often use blocks as the finish for a wall.

When you find a face block wall, make a note of the:

- type of block – is it textured or smooth?
- neatness of any cut blocks – do you think the cut edges are accurate enough?
- finish applied to the mortar joint – is it neatly produced or does it spoil the finish?

Lightweight insulation blocks are manufactured from a softer material than concrete blocks. This means that, while they can be cut using a club hammer and bolster, they can also be cut using a hand masonry saw.

▲ Figure 3.31 Cutting lightweight insulation blocks with a hand masonry saw

Accurate cuts can be achieved easily using a saw, and more difficult shapes can be produced using a brick or scutch hammer. This is useful when you need to form a channel or hole, for example, to allow pipes or cables to be installed.

It takes practice and experience to be able to cut blocks by hand accurately and efficiently.

INDUSTRY TIP

Hand masonry saws can be expensive to buy because they have teeth that are tipped in tungsten (a hard metal material that lasts a long time). Keep a lookout for carpenters on site who often dispose of hand saws that have become too blunt for accurate carpentry work. These discarded saws are usually fine for cutting insulation blocks until they lose their sharpness completely.

If you need to cut a large quantity of blocks to size, you will be able to work more quickly and reduce waste if you use a powered masonry saw. This could be a handheld power tool or a bench mounted machine (sometimes referred to as a clipper saw).

Whether block cutting is done by hand or machine, you should aim to produce accurate cuts as quickly as possible.

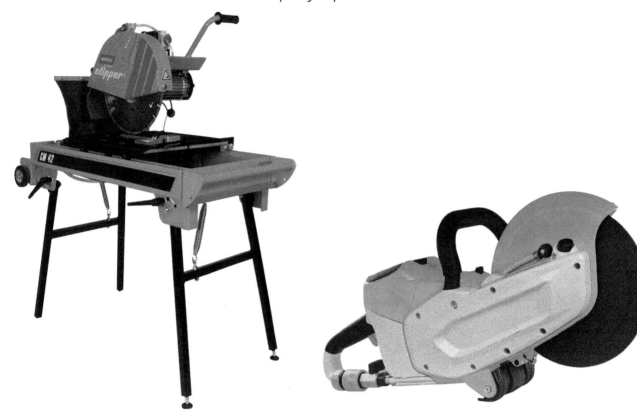

▲ Figure 3.32 Masonry bench saw/clipper saw

▲ Figure 3.33 Disc cutter (a handheld powered masonry saw)

HEALTH AND SAFETY
Never use powered cutting equipment unless you have been properly trained and certified as competent.

IMPROVE YOUR ENGLISH

Search online for 'block splitter'. Copy and paste an image into a Word document and write a description of how this piece of equipment cuts blocks accurately without power.

3.4 Forming mortar bed joints

The techniques required to manage and manipulate the materials used in constructing masonry using blocks (and bricks) will be repeated over and over throughout a bricklayer's career. This is especially true with the methods used to produce bed and perp joints in mortar.

First, we will consider how a mortar bed joint is formed. Rolling mortar on a spot board is the traditional way of preparing the material to lay a bed joint. This process applies to both blocklaying and bricklaying.

From the shaped mound of mortar on the spot board, a portion of mortar is sliced off using the blade of a trowel held vertically. A roll is formed by a back-and-forth horizontal movement across the board. Keep in mind that the rolling action takes a great deal of practice to master.

Study the step-by-step guide to see how this works.

Step by step
Rolling mortar to form a bed joint

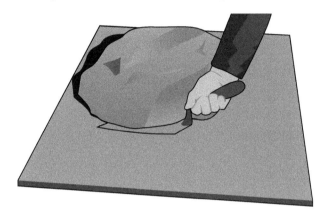

Step 1 Form the mortar into a neat mound on the spot board. Leave an area next to it where the mortar can be rolled.

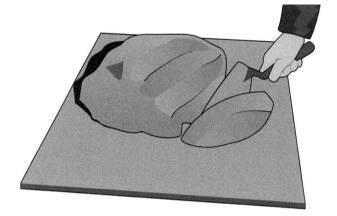

Step 2 Slice off a portion of mortar. Do not slice off too much – just enough to handle comfortably. The amount you can work with will increase with practice.

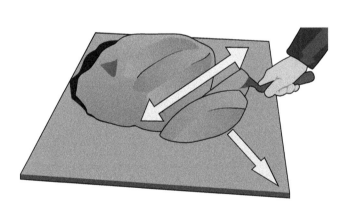

Step 3 With a back-and-forth motion, push the mortar across the spot board to create a sausage shape. Pick up the mortar and place it back at the start position to roll it again. It may take many repetitions to get this right, depending on the workability of the mortar.

Step 4 When the mortar roll is evenly formed, pick it up from the spot board with a sweeping action to maintain its shape. It is now ready to 'throw' as a bed joint along the line of the wall.

INDUSTRY TIP

With experience, you will become faster at preparing the mortar to lay a bed joint. While rolling is the traditional method of preparing the mortar on a spot board, many bricklayers develop the ability to use the underside of the trowel blade to 'pat' the mortar into shape ready for spreading.

Accurately placing the rolled mortar on the wall to make a bed joint is another skilled process that requires a great deal of practice. The prepared mortar is placed on the wall with a swift action that controls the flow of mortar into the desired position. Once the mortar is in position, it can be distributed as necessary along the length of the wall by a spreading action with the back and tip of the trowel blade.

Study the step-by-step guide to see how this works.

Step by step
Laying a mortar bed joint

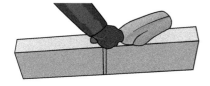

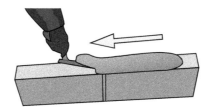

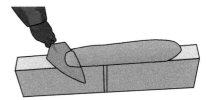

Step 1 Carefully position the loaded trowel of mortar above where the bed joint is to be laid.

Step 2 With a swift but controlled sweeping motion along the line of the wall, make the mortar flow into the desired position.

Step 3 Make any adjustments needed to prevent droppings falling from the wall. If the bed joint is thicker or thinner in places, even it out.

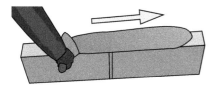

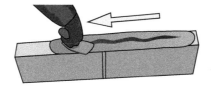

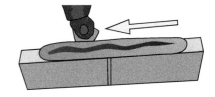

Step 4 With the underside of the trowel blade, gently smooth out the top of the mortar. Take care not to cause mortar to squeeze over the edge of the wall and fall off.

Step 5 Use the tip of the trowel to create a small groove along the top of the mortar. This will help the block (or brick) to settle into position more easily. Notice the angle at which the trowel is held in relation to the wall.

Step 6 Finally, run the edge of the trowel along the back edge of the bed joint to create a shallow angle (called a chamfer). This will reduce the amount of mortar squeezing out of the bed joint when a block is placed in position.

INDUSTRY TIP

When carrying out Step 5, be careful not to create a deep furrow, as this could produce small pockets under the blocks or bricks when the mortar hardens. Water could stand in these pockets and cause frost damage during freezing weather conditions.

With practice, you will be able to accurately judge the thickness of the mortar, so that when a block (or brick) is placed on the bed joint it will not need to be struck repeatedly to position it, and there will be only a small amount of surplus mortar squeezed from the joint.

3.5 Forming mortar perp joints

Producing a mortar perp joint requires the use of practised techniques, because the process often means working against gravity – the mortar joint can easily fall off the end of the block as it is being positioned. We will look at two ways of producing perp joints.

Study the step-by-step guide to see how to produce a perp joint on a block already laid as part of the wall.

Step by step
Perp joint method 1

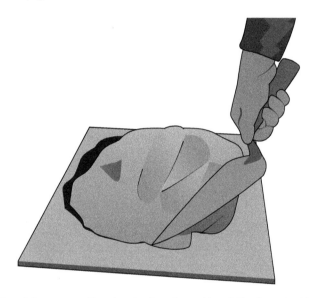

Step 1 Prepare a portion of mortar from the spot board. Use the underside of the trowel blade to repeatedly pat the mortar and make a small sausage shape about the size of the end of the block.

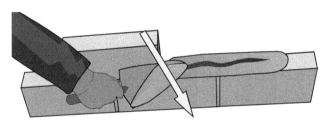

Step 2 'Settle' the shaped mortar onto the trowel blade by firmly shaking the trowel downwards. This will help to prevent the mortar falling from the trowel when forming the perp.

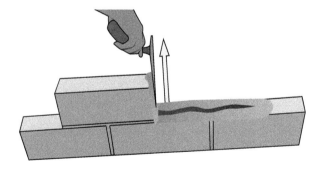

Step 3 With a smooth action, press the shaped mortar into position on the end of the block. While maintaining the pressure on the block, slide the trowel blade carefully upwards. As it reaches the top edge of the perp, increase the pressure slightly in a flicking motion before lifting the trowel quickly away from the perp.

Study the next step-by-step guide to see how to produce a perp joint on a block before it is laid as part of the wall. The block is stood on end and the perp is formed in a similar way to the method used to form a bed joint.

Step by step
Perp joint method 2

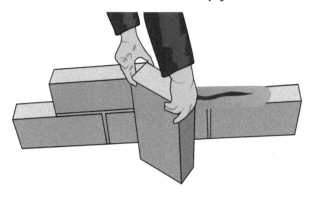

Step 1 Stand the block upright on end. Make sure it is stable when positioned.

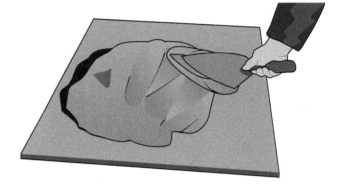

Step 2 Use the underside of the trowel to form a portion of mortar on the spot board to match the end shape of the block.

Step 3 With a flowing action, place the shaped mortar onto the end of the block. Smooth the surface of the mortar and 'wipe' the trowel blade along the edges of the block end to make the mortar stick securely.

Step 4 Lift the block carefully into position and squeeze the mortar perp joint tight against the block previously laid.

Over time, bricklayers develop their own style when using the techniques described here. With practice and persistence, levels of productivity and accuracy will improve.

3.6 Building block quoins

In Chapter 5, there are step-by-step instructions on building quoins (corners) using bricks and blocks to construct a cavity wall. Those steps also apply to building a single-leaf wall using blocks, and the sequence of work is discussed in this section.

Usually, a **return** quoin or corner is built at a 90° angle (or 'right angle'). It is important to create this angle accurately when setting out a quoin. This is known as maintaining 'square' and can be achieved by using a builder's square.

In the case of brick quoins, it is easy to maintain half bond throughout the return because the dimensions of bricks are modular (see Chapter 4 for an explanation of what this means). To maintain half bond in a block quoin, a cut block must be introduced next to the quoin block in each course. Figure 3.35 shows where a 100 mm cut block is positioned to maintain half bond in the return.

> **KEY TERM**
>
> **Return:** expression used in bricklaying to describe the portion of blockwork (or brickwork) at right angles to the face of the wall

INDUSTRY TIP

Some bricklayers use a brick placed vertically on end against the quoin block with thicker perp joints, instead of a cut block, to create half bond in a block corner. This is not good practice. If the brick is not the same material as the blocks, it will respond to temperature changes differently, which could lead to problems later.

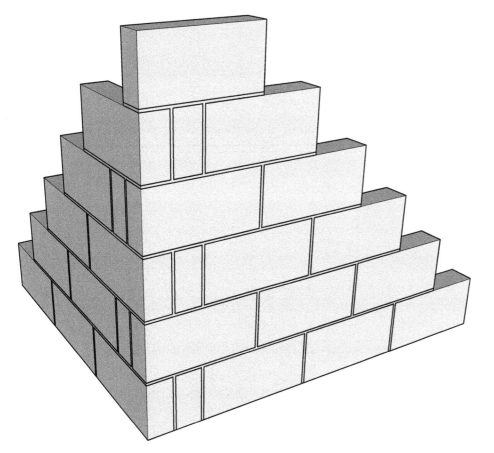

▲ Figure 3.34 Right-angled block quoin

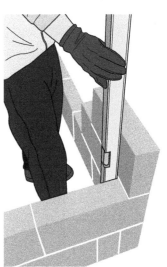

▲ Figure 3.35 Plumbing an internal block quoin

Since most building designs are square or rectangular in shape, building corners that are square is crucial to achieving overall accuracy when setting out.

Keep in mind that quoins in walls can be external or internal. Whichever type of quoin is built, the same skills and checks apply to the stages of construction.

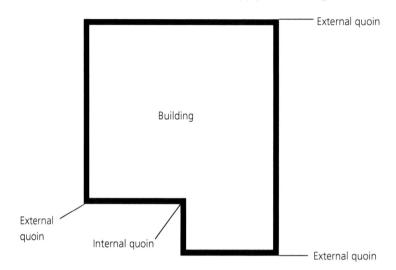

External quoin

Building

External quoin

External quoin

Internal quoin

External quoin

▲ Figure 3.36 External and internal quoin positions in a building

ACTIVITY

Draw a simple plan view sketch of an H-shaped building (use a ruler for neatness). How many internal quoins will be constructed?

If the quoin is in the form of a return, or the wall has stopped ends, each course will be 'racked back'. This means that the blocks will form steps as the courses are laid.

▲ Figure 3.37 Blockwork 'racking back'

As each course in the quoin is laid, the work must be gauged, levelled, plumbed and ranged to ensure accuracy. Study the step-by-step guide to see how this is done.

Step by step
Laying courses in the quoin

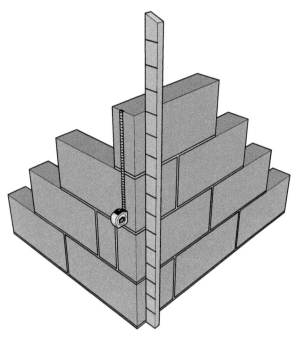

Step 1 Gauge each course with a gauge rod or tape measure.

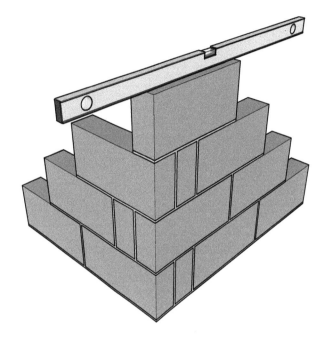

Step 2 Level each course with a spirit level.

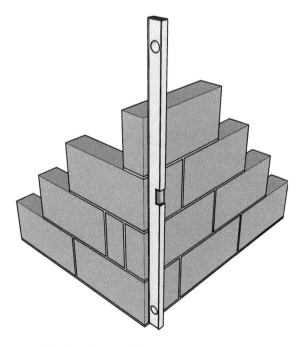

Step 3 Plumb each course with a spirit level.

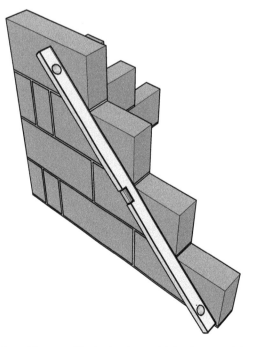

Step 4 Range racking back using a spirit level as a straight edge.

Once quoins or stopped ends for a wall have been completed accurately, a string line can be attached to build the infill between them.

3.7 Laying to the line

Laying to the line is another skill that the bricklayer develops over time. Speed and accuracy only come with practice and experience.

As mentioned previously in this chapter, the bricklayer must practise accurately forming the mortar bed joint to the correct thickness. This prevents the need to strike the block excessively to follow the guide provided by the string line.

Keep in mind the following points when developing and improving your skills:

- Make sure the top **arris** of the block follows the string line along its full length.
- Never lay the blocks touching the line. Doing this will cause the line to move away from the face of the wall, so an accurate **face plane** will not be achieved.
- 'Eye' down the wall to check the face plane is even and the perps are lined up vertically.

When building a single-leaf wall using block, take care not to build too high. Be aware that the wall will be vulnerable to the effects of wind gusts even after the mortar has hardened.

Official guidance provided by **British Standards** (BS EN 1996-3:2006) states that a single-leaf wall built using block should not be raised more than six courses in one operation.

Manufacturer's guidelines also give information on safe practice for building heights of block walls under different working conditions. These guidelines, along with the relevant British Standards, are often included in method statements.

▲ Figure 3.38 'Eyeing' down the wall

KEY TERMS

Arris: the long, straight, sharp edges of a block or brick formed at the junction of two faces

Face plane: the accurate alignment of all the blocks or bricks in the face of a wall to give a uniform flat appearance

British Standards: the UK authority that sets out a range of standardised quality requirements, procedures and terminology

3.8 Building block junctions

The design of a structure will often include locations where one wall forms a junction with another wall. A junction is formed by bonding the masonry in a specified way to create a stable and strong feature.

The traditional method of bonding block courses at a junction requires the use of cut blocks where the walls meet. This can lead to a lot of cutting work, which will often produce waste.

Sometimes the junction wall is built after the main wall is constructed. This is achieved by leaving pockets (or 'indents') in alternate courses of the main wall following the vertical line of the junction wall. The corresponding courses of the junction wall can then be built into the indents in the main wall later, to complete the junction.

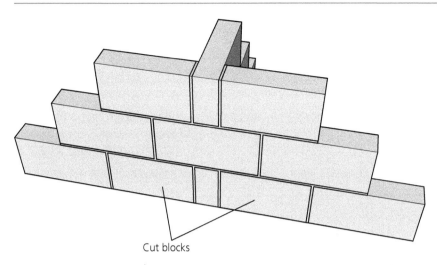

Cut blocks

▲ Figure 3.39 Junction in a single-leaf block wall (note the use of cut blocks to form the bond)

Other more modern methods of constructing a junction are quicker and less wasteful, making use of components that reduce the need to cut blocks.

A relatively simple method is to use expanded metal lath (EML), which is a type of reinforcing mesh. This is carefully built into the bed joints on the vertical line of the wall that will be added later. The mesh is built in as work on the main wall proceeds, in accordance with the specifications.

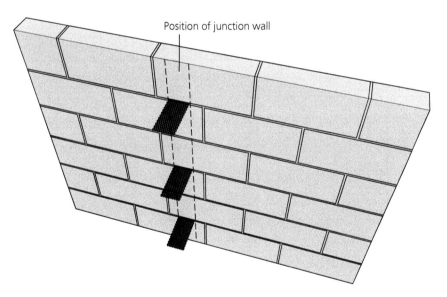

Position of junction wall

▲ Figure 3.40 Reinforcing mesh built into bed joints to receive a block junction wall later

A more modern approach to forming a stable junction between walls is to use **proprietary** connectors or wall-starter kits, which are efficient and easy to use.

IMPROVE YOUR ENGLISH

Visit www.catnic.com and use the search term 'stronghold' to find the page on wall starter kits.

Study the datasheet and write down the benefits of using this system to form junctions in masonry.

> **HEALTH AND SAFETY**
> Reinforcing mesh has sharp edges. If it is left projecting from the wall ready to receive a junction wall later, it should be folded down to keep it out of the way until it is needed.

> **KEY TERM**
> **Proprietary:** manufactured and sold under a brand name or trademark

3.9 Forming joint finishes

A block wall is sometimes specified as a face wall, meaning it will serve as the finish that will be on show for the life of the structure. This will require a joint finish to be formed for the bed and perp joints.

Forming joint finishes (or 'jointing') requires patience and practice to get a finish that is visually acceptable. Using a jointer or jointing iron will smooth and compress the surface of the mortar to seal it and produce a half-round profile to the joints (often referred to as concave, Figure 3.41). The main consideration is to achieve a neat, attractive finish.

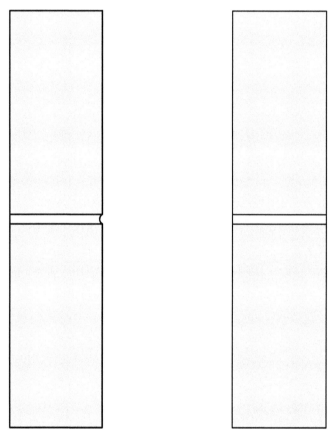

▲ Figure 3.41 Blockwork concave or half-round joint

▲ Figure 3.42 Blockwork flush joint

If the blockwork is to be covered over and out of sight, the main consideration is to make sure that the mortar joints are free from gaps or small openings that could form weak points in the wall. The finish will be a flush joint (Figure 3.42) produced using either the underside of the laying trowel or a pointing trowel.

Test your knowledge

1 Which document shows a list of repeating components or features and the site locations where they are to be installed?

   a Specification

   b Method statement

   c Risk assessment

   d Schedule

2 If a wall is 2 m long, how long will it be on a scale drawing with a scale of 1:10?

   a 100 mm

   b 200 mm

   c 210 mm

   d 220 mm

3 Which type of drawing shows a slice through a structure to reveal details that would otherwise be hidden?

   a Plan

   b Hatching

   c Section

   d Elevation

4 What is the length in millimetres of a standard block?

   a 140

   b 240

   c 340

   d 440

5 What must be added to pre-mixed mortar to ensure it remains workable throughout a working day?

   a Restrictor

   b Resealer

   c Retarder

   d Repeater

6 When the end of a wall is formed without a return quoin (or corner), how is it described?

   a Stepped end

   b Sloped end

   c Stopped end

   d Staggered end

7 When building quoins using blocks, what size of cut is used at the corner to maintain half bond?

   a 75 mm

   b 100 mm

   c 120 mm

   d 200 mm

8 To follow British Standards, what is the maximum number of courses a block wall should be raised in one operation?

   a 6

   b 8

   c 10

   d 12

9 Which letters refer to reinforcing mesh?

   a ELN

   b EML

   c ENL

   d ELM

10 What is another description for a concave joint finish?

   a Half-formed

   b Flush

   c Half-round

   d Flat

CHAPTER 4

CARRYING OUT BRICKLAYING ACTIVITIES

INTRODUCTION

The skills and work methods used for blocklaying activities (covered in Chapter 3) can also be applied to bricklaying activities, such as rolling mortar to form a bed joint. However, laying bricks can require greater precision and care, especially if the brickwork forms the face of a building which will be on show for many years.

This chapter considers the preparation and specific skills needed to carry out bricklaying activities. It also details the tools and equipment required to set out and build straight walls, corners and junctions in brick.

LEARNING OUTCOMES

After reading this chapter, you should:
1 know how to prepare for bricklaying activities
2 know how to identify and select the correct resources for bricklaying activities
3 know how to build straight walls, corners and junctions using bricks.

1 PREPARATION FOR BRICKLAYING ACTIVITIES

Safe working practices are vital during preparation for any construction task. As mentioned in previous chapters, there are many laws and regulations that must be followed to protect everyone on site. These are detailed in Chapter 6 and should be reviewed often.

1.1 Dealing with common hazards

Remember, everyone on a construction site has a responsibility to make sure they work in a safe manner, to protect themselves and those around them. We will review some common hazards that a bricklayer might encounter when preparing for a work task.

Hazardous materials

There are materials used in construction that must be handled with care. For example, cement powder used in mortar can cause skin irritations such as dermatitis, or it can even cause chemical burns if the skin is in contact with the mortar mix for long periods.

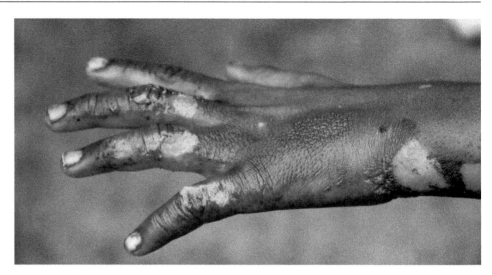

▲ Figure 4.1 A chemical burn as a result of not wearing safety gloves

Always follow the manufacturer's directions and technical guidance on how to use products safely. This includes wearing appropriate personal protective equipment (PPE). In Chapter 6, there is important information about regulations, including the Control of Substances Hazardous to Health (COSHH) Regulations 2002.

Mixing mortar

Mixing mortar may require the use of machinery, such as a drum mixer. Drum mixers are available in a range of sizes, according to the quantity of mortar required for the job. Any size of powered mixer must be treated with respect, since carelessness when mixing could lead to severe injury.

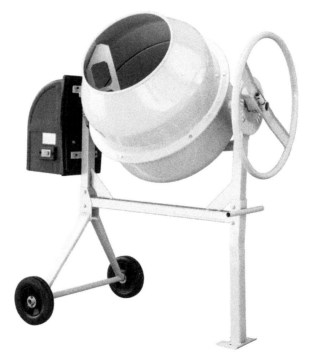

▲ Figure 4.2 Drum mixer

Moving and stacking materials

Bricklaying materials are often heavy and awkward to move and handle. When stacking bricks and lifting materials, potential hazards include crushing injuries, muscle strains, and cuts and bruises to exposed areas of skin. Plan ways to minimise manual handling and make good use of mechanical handling provisions on site such as forklifts and cranes.

> **INDUSTRY TIP**
>
> If heavy materials or components can only be moved manually, a sack truck or pallet truck can be used if the ground is level and firm enough.

Cutting bricks

A bricklayer often needs to cut bricks to length or trim them to shape using the appropriate hand tools (the correct methods of cutting bricks are discussed in detail later in this chapter). Cutting operations are potentially hazardous because they create dust particles and flying brick fragments that can cause eye injuries.

▲ Figure 4.3 Forklift

Using hammers with chisels can be hazardous; there is the potential for operatives to suffer hand injuries when applying strong hammer blows during brick cutting.

The end of a steel chisel (sometimes referred to as 'cold chisel') can change shape when struck repeatedly to form what is often referred to as a 'mushroom head'. This can produce sharp edges which can cause hand injuries and the sharp projections should be removed when necessary with a suitable grinding tool.

▲ Figure 4.4 Cutting a brick

> **HEALTH AND SAFETY**
>
> Using the correct PPE is essential for all bricklaying activities, especially during the preparation stage. Moving materials to the work area and positioning them will expose workers to hazards that can be minimised by using the correct PPE, such as safety boots, a hard hat, hi-vis clothing, gloves, safety glasses and a dust mask or RPE (respiratory protective equipment).

Through careful planning and preparation for work tasks, it is possible to reduce the risk of harm occurring. Documents such as risk assessments and method statements provide important information to help you work safely. Make it a habit to consult these documents before starting work.

▲ Figure 4.5 Consulting a risk assessment

1.2 Information sources when preparing

It is important to refer to a range of information documents in order to understand all the design details of a work task.

A key document is the specification, which gives information about the type of bricks and other materials to be used. It may also provide details on the work practices that must be employed for a specific job, to achieve the required quality of finish. An example specification is shown in Figure 4.7.

When a component or feature of a work task is used for several jobs on site, it may be listed in a schedule. For example, this could be a decorative design feature that is repeated in different wall locations around the site, perhaps using different coloured bricks in different places.

▲ Figure 4.6 Checking information documents

ACTIVITY

Create a schedule in the form of a simple table to show decorative brick features on each wall of a rectangular building. Label the walls 1–4 and identify a brick for each decorative feature.

You could go to a brick manufacturer's website and choose different named coloured bricks from their range to include in your schedule.

TECHNICAL SPECIFICATION

Brick Type				New Products (higher specification)		
Brick Type	**HANDMADE**	**TRADITIONAL WIRECUT**	**RECLAIM & CHERWELL**			
Brick Codes				CHB	VGB (Wirecut)	Cotswold Collection
Appearance	Lightly creased genuine handmade	Extruded wirecut smooth, rustic, plain or sandfaced	Prematurely 'aged' during the production process			
Specification	Bricks are manufactured to BS EN 771-1 (BS 3921:1985 now withdrawn)					
Sizes	Metric: 215mm x 102.5mm x 50mm, 65mm 68mm 73mm 80mm Imperial: 9" x 4⁵/₁₆" x 2", 2¼", 2³/₈", 2½", 2⁵/₈", 2⁷/₈", 3", 3¹/₈"		Non Standard: Most sizes can be accommodated in our production process			65mm 73mm
Specials	A full range of standard BS and purpose made Specials is available					
Compressive Strength	>24 N/mm²	>60 N/mm²	>60 N/mm²	>60 N/mm²	>60 N/mm²	>60 N/mm²
Durability	F2	F2	F2	F2	F2	F2
Tolerances	T1	T2	T1	T1	T1	T1
Range	R1	R1	R1	R1	R1	R1
Soluble Salt Content	S2	S2	S2	S2	S2	S2
Water Absorption	<17 %	<12 %	<12 %	<12 %		
Packaging	Shrinkwrapped and wire banded with holes for forklift use					
Pack Sizes & Weights	Size (mm) / Qty/pack / Wt (t) 50 650 1.229 65 500 1.223 73 500 1.280 80 400 1.213	Size (mm) / Qty/pack / Wt (t) 50 650 1.070 65 500 1.072 73 500 1.195 80 400 1.100	Size (mm) / Qty/pack / Wt (t) 50 650 0.927 65 435 0.929 73 400 1.009 80 320		Size (mm) / Qty/pack / Wt (t) 50 650 1.070 65 500 1.072 73 500 1.195 80 400 1.100	
Workmanship	The recommendations of good practice made in the relevant British or European Standards regarding design and workmanship must be fully observed. Always mix from at least 3 packs, working diagonally down the blades, not horizontally across the tops of packs. Bricks & brickwork should be covered during construction to prevent saturation.					
Performance	All Northcot 1st quality bricks are Frost Resistant. Northcot Bricks are manufactured to F2 rating for Frost Resistance as described in BS EN 771-1: 2011 "Specification for Clay masonry units". The specification is for masonry (walling) products which are to be used in brickwork designed and built in accordance with recommendations in BS PD 6697: 2010 "Recommendations for the design of masonry structures to BS EN 1996-1-1 and BS EN 1996-2". The use of Sulphate Resisting Cement is recommended. Further details on website					
Samples	Usually within 48-hour dispatch					
Important Advice	Colours and textures reproduced here are as accurate as the printing process allows and final choices should not be made from this brochure in isolation. For the latest updates on technical information, please visit our website on www.northcotbrick.co.uk					

▲ Figure 4.7 A specification (note the amount of detail it contains about brick types)

Other documents used when preparing for construction of a brick wall include:

- job sheets
- manufacturer's instructions
- written safety policies and procedures.

To make sure information is presented to the bricklayer in a clear and understandable way, printed text documents are used together with a range of working drawings.

1.3 Drawings as information sources

Drawings are an information source that can provide a bricklayer with a lot of clear information without the need for lots of writing.

In this chapter, we will focus on three types of drawing:

- plan
- elevation
- section.

▲ Figure 4.8 Drawings provide a lot of information in a clear way

Most drawings show a scale representation of a work task. If a drawing of a brick wall were drawn full size, the document would be almost impossible to use. As mentioned in Chapter 3, scale is shown using a ratio such as 1 to 10 (usually written 1:10).

INDUSTRY TIP

When reading drawings, always work to the written dimensions. Although it is possible to take dimensions off a scale drawing using a scale rule, they should not be relied on as accurate.

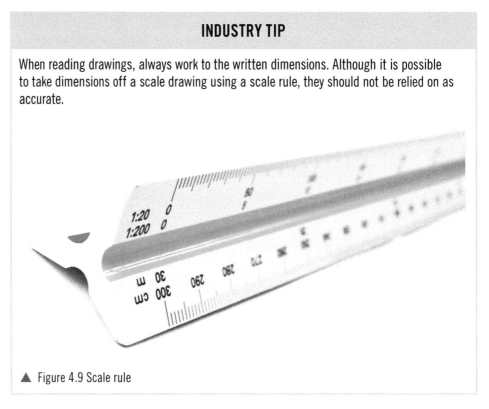

▲ Figure 4.9 Scale rule

Study the illustrations in Table 4.1, which show the different types of drawing in relation to a half-brick-thick wall.

▼ Table 4.1 Types of drawing for a half-brick-thick wall

Type of drawing	Description	Example
Plan	A view from above the work task (think of it as a bird's-eye view) Scale 1:50 or 1:100	Plan view of a half-brick-thick wall 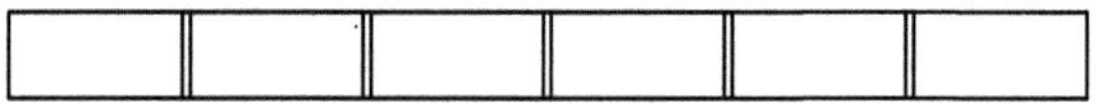
Elevation	A view showing each face or side of a work task Scale 1:50 or 1:100	Front elevation of a half-brick-thick wall 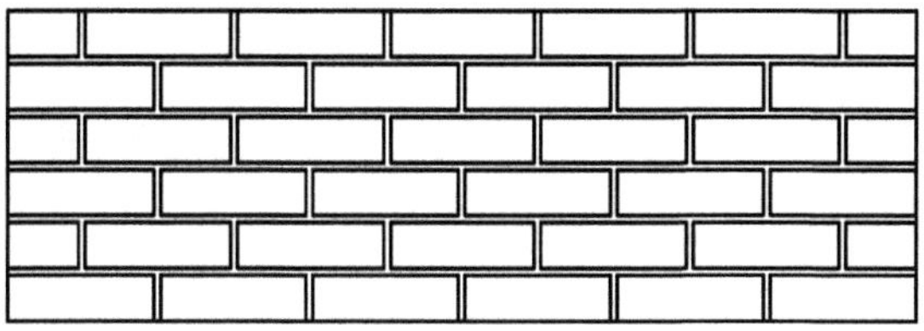
Section	A slice through a work task, which can show details that would otherwise be hidden Scale 1:50 or 1:100	Section view of a half-brick-thick wall 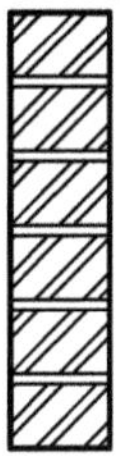

The diagonal lines in the section drawing in Table 4.1 show that the material within the 'slice' of the wall is brick. Remember that lines and symbols used to show different materials are referred to as hatchings.

ACTIVITY

Draw your own copy of the section drawing in Figure 4.10 and fill in the hatchings that apply to each described material. Refer to Figure 1.8 in Chapter 1 to help you.

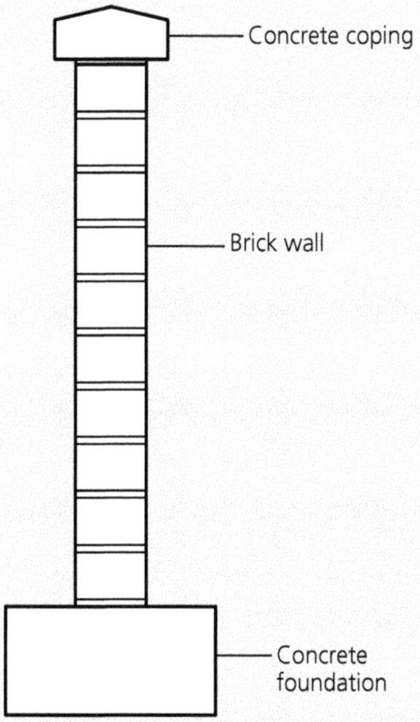

INDUSTRY TIP

If greater detail is required for specific parts of a drawing, a scale of 1:10 can be used to allow a larger image to be produced.

Concrete coping

Brick wall

Concrete foundation

▲ Figure 4.10 Section drawing without hatchings

② IDENTIFYING AND SELECTING THE CORRECT RESOURCES FOR BRICKLAYING ACTIVITIES

It is important to make lists of resources before starting to build, so you can calculate accurately the required quantities of materials. This supports efficiency; if sufficient quantities of materials are not available at the work location, time could be wasted obtaining more when they run out.

2.1 Calculating quantities

As mentioned in previous chapters, calculations of volume and area are frequently used when working out required quantities of construction materials. In this chapter on bricklaying activities, the focus is on calculating area.

Remember, to calculate the area of the face of a wall, multiply the length by the height. Remember, the length and height dimensions along straight lines are known as 'linear' measurements.

The surface area of a wall is calculated in m² (square metres). Each m² of a half-brick-thick wall contains 60 bricks. Therefore, to calculate the number of bricks required to build a wall, you simply multiply the number of m² by 60.

It is difficult to avoid generating some waste when performing bricklaying activities:

- Bricks may be damaged when they are moved from storage to the work location.
- When bricks are cut, offcuts cannot always be used.
- Some bricks may have manufacturing faults.

It is therefore important to add a percentage for waste when calculating quantities.

INDUSTRY TIP

The amount of allowance for waste will vary, depending on several factors:
- Some bricks are soft and damage easily.
- Other bricks may be harder but are brittle and break easily.
- The task may involve cutting a lot of bricks to size.

As a general rule, add 5% wastage to the quantities calculated, but be prepared to adjust this amount according to circumstances.

IMPROVE YOUR MATHS

Try working with someone else to calculate area. Take turns to invent wall dimensions that are not whole metres and calculate the area in m². Keep the dimensions sensible – up to a maximum of 10 m long and 2 m high. Check each other's work and discuss any errors that you make.

2.2 Checking suitability

A bricklayer should carefully check the condition and suitability of materials and resources before using them to build a wall, to maintain high standards of work. For example, if damaged bricks are used, this can affect the appearance of the finished wall, as well as potentially reducing its strength.

Bricks are delivered to site in packs which usually include some form of weather protection such as plastic wrapping. Make sure you think ahead about maintaining protection of bricks and other materials. Having polythene sheeting or other protection on hand is good practice at the preparation stage.

Materials should be protected from damage caused by bad weather at all times. Bricks are difficult to lay accurately and neatly when wet. Equally importantly, problems such as frost damage can emerge later if bricks are wet when they are laid. (You will learn much more at Level 2 about the potential damaging effects of moisture and the chemical reactions that can occur when bricks are saturated.)

When storing bricks and transporting them from storage to the work location, always handle them with care, to minimise waste and to ensure the appearance of the finished wall meets high standards.

▲ Figure 4.11 Bricks protected by polythene sheeting

2.3 Range of materials

Bricks

Bricks are manufactured in a vast range of colours and textures. They also vary in hardness or strength and resistance to moisture penetration, according to the materials used in manufacture. Knowing the characteristics of the materials selected is important when planning how to store and move them. There is much more about the characteristics of bricks, especially when used to construct cavity walls, in Chapter 5.

Bricks fall into three main categories:

- facing bricks
- engineering bricks
- common bricks.

Facing bricks

Facing bricks can form the face of a wall. The colour and texture specified will create the decorative appearance of the finished wall.

▲ Figure 4.12 Facing bricks

Engineering bricks

Engineering bricks are strong and can withstand high loadings. This means that they can resist forces that squeeze or stress them. They do not absorb moisture, so they are suitable for sections of a wall below ground level. They can also be used as a DPC (damp-proof course) in a wall, to prevent moisture being absorbed upwards from saturated ground.

Common bricks

Common bricks are lower quality bricks often used in locations where they will not be seen, such as a wall that will be coated (or **rendered**) with sand/cement mortar.

▲ Figure 4.13 Two courses of engineering brick can form a DPC

KEY TERM

Rendered: where a wall is coated with a sand/cement mix to provide a smooth, weatherproof finish

ACTIVITY

Visit the websites of some brick manufacturers (try searching for Ibstock Brick as a starting point). They will usually have a section on the materials they use for manufacturing their bricks. List the names of three manufacturers you have found and what materials they use.

▲ Figure 4.14 A common brick with frog (the depression in the top of the brick)

Mortar

Mortar for bricklaying activities is similar to the mortar used for blockwork, as detailed in Chapter 3. The main mortar materials are sand, cement and water. The consistency of the mortar will vary, depending on the type of brick being used. For example, a common brick is more absorbent than an engineering brick, so the mortar will be mixed to a different consistency.

▲ Figure 4.15 Getting the right mortar consistency takes practice

To make the mortar more workable, a plasticiser is added during the mixing process, which must always be used in accordance with the manufacturer's instructions. Using too much plasticiser can affect the strength of the mortar when it hardens.

IMPROVE YOUR ENGLISH

Review the details in Chapter 3 about mortar. In your own words, write a short explanation of the way lime powder acts as a plasticiser compared to a liquid chemical plasticiser.

▲ Figure 4.16 Plasticiser

2.4 Range of tools

Before starting work, make a list of the tools and equipment needed for the job. This is good practice and contributes to efficiency. Remember, the tools needed for any masonry task can be split into three main groups:

- laying and finishing
- checking
- cutting.

Table 4.2 lists the tools required for building walls using bricks. Take care of your tools and always check they are in a safe condition before using them.

▼ Table 4.2 Tools used in bricklaying

Category	Tool	Description
Laying and finishing	Trowel	Used for: ● forming and laying mortar bed joints ● forming and buttering perp joints
	Pointing trowel	Used for shaping mortar joints Can be used for smoothing mortar applied to the top of sections of masonry
	Jointer	Used for forming a joint finish to bed and perp joints in face work, a process sometimes called 'ironing'
Checking	Tape measure	Used to set out and check dimensions
	Spirit level	Used to check work is plumb and level
	Line and pins	Used as a guide to accurately lay bricks along the length of a straight wall

▼ Table 4.2 Tools used in bricklaying (continued)

Category	Tool	Description
Checking	Corner block Corner block	Used to position a line and pins on corners (or quoins)
Cutting	Club hammer	Used together with a brick bolster to accurately cut bricks to the required dimensions Sometimes called a lump hammer .
	Brick bolster	Used like a chisel with a club hammer to cut bricks accurately to the required dimensions
	Brick hammer	Used to cut and trim bricks quickly where accuracy is not critical
	Scutch hammer	Used to accurately and finely trim and shape bricks as required

2.5 Range of equipment

Table 4.3 shows some of the equipment used for bricklaying activities.

▼ Table 4.3 Equipment used in bricklaying

Equipment	Description/How it is used
Spot boards 	Boards that have mortar placed on them within easy reach of the bricklayer along the length of a wall
Shovel 	Used to move mortar and loose materials or for mixing mortar by hand
Wheelbarrow 	Essential for moving materials to the work location
Brick tongs 	A specialist piece of equipment used for efficiently and safely picking up a number of bricks at once

▼ Table 4.3 Equipment used in bricklaying (continued)

Equipment	Description/How it is used
Gauge rod 525 mm 450 mm 375 mm 300 mm 225 mm 150 mm 75 mm	Used to check that brickwork is constructed to accurate height measurements with consistent joint sizes Sometimes referred to as a 'lath rod' A variation known as a 'storey rod' is used to check important height levels, such as the tops of doors and windows
Builder's square	Used for accurately setting out work with right-angled corners
Straight edge	Used to transfer levels over distances longer than a spirit level
Tingle plate Tingle plate	Used to support a string line stretched over long distances (a long string line could sag in the middle; if a wall is built following the line, it will also sag in the middle)

To complete the list of tools and equipment, we can add buckets, sweeping brushes and hand brushes. A soft hand brush is useful for lightly removing mortar particles from the face of brickwork to produce a clean finish.

INDUSTRY TIP

Some bricklayers prefer not to brush the face of a brick wall to produce the final finish. If you do decide to use a soft brush, take great care not to create smudges on the face of the bricks by brushing when the mortar is still soft. Use a light touch.

HEALTH AND SAFETY

Remember, maintaining and cleaning your tools properly contributes to health and safety and will make them last longer.

3 BUILDING STRAIGHT WALLS, CORNERS AND JUNCTIONS IN BRICK

After completing the initial preparation stage and identifying resources, the work area can be set out for the work task. Remember, success depends on careful preparation.

Important points regarding positioning materials and components ready for work are covered in detail in Chapter 5, where consideration is given to methods of safely and efficiently moving and stacking both bricks and blocks when building cavity walls.

In this chapter, we will concentrate on how to set out brickwork in a straight, half-brick-thick wall. We will also discuss how to set out and bond corners and junctions.

3.1 Establishing the bond

The most common brick bond is stretcher bond, which is often referred to as 'half bond'. (The long face of a brick is called a 'stretcher'; Figure 4.20 shows the named parts of a brick.) This is because each alternate course of bricks is laid with an overlap of half the length of a standard brick. This means that the perp joints in any one course are exactly halfway along the stretcher faces of the bricks in the course below. (See Section 3.3 on pages 123–5 for more details on perp joints.)

▲ Figure 4.17 Stretcher (or half) bond

The overlap of bricks in alternate courses adds to the strength of the wall. Loadings on the wall, such as a beam pressing down on it, are evenly distributed through the wall to reduce the risk of potential long-term damage. Study Figures 4.18 and 4.19 to see the effect of bonding a brick wall correctly to support loadings.

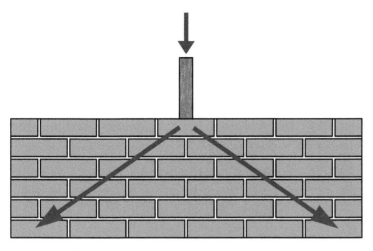

▲ Figure 4.18 Loadings spread through a correctly bonded wall

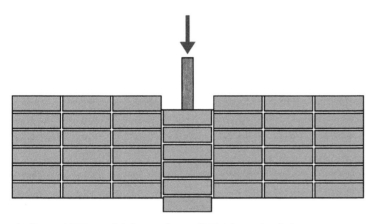

▲ Figure 4.19 Potential damage in a wall that is not bonded

INDUSTRY TIP

There are other bonding arrangements that are referred to as 'quarter bond'. Typical quarter bonds are English bond and Flemish bond, which are covered in greater detail at Level 2.

Bricks are manufactured to dimensions and **tolerances** decided by official institutes. In the UK, the British Standards Institute is the agency that produces official standards and gives numbers to components so that details about them can be checked easily.

For many years, the European Union has influenced standards, with the result that many numbers have been changed. For example, clay bricks used to have the number BS 3921. This was changed to BS EN 771-1. Further changes are likely with the UK leaving the EU.

Study Figure 4.20 to become familiar with the dimensions of a brick and the names of its various parts. In time, you should know these off by heart. This is important information to have in mind when setting out the bond.

KEY TERM

Tolerances: allowable variations between specified measurements and actual measurements

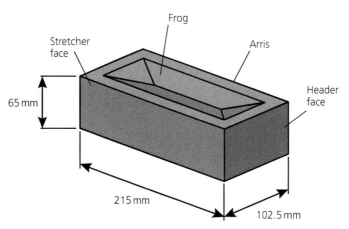

▲ Figure 4.20 Dimensions and parts of a brick

Note the width of a brick: 102.5 mm. Why bother with 0.5 (half) a millimetre? It is important because bricks are designed to be assembled in **modular** patterns, which means the dimensions of each face or side of the brick can be connected or combined in different ways. See how this works in Figure 4.21.

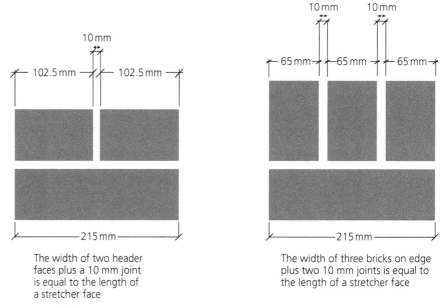

The width of two header faces plus a 10 mm joint is equal to the length of a stretcher face

The width of three bricks on edge plus two 10 mm joints is equal to the length of a stretcher face

▲ Figure 4.21 How the dimensions of a brick are modular

Due to variations in the quality of materials and in manufacturing processes, bricks can vary slightly in overall size. We can accommodate these differences by adjusting the size of the perp joints between bricks.

There is an allowable tolerance for perp joints of plus or minus 3 mm. This means that the maximum joint size should be 13 mm and the minimum joint size should be 7 mm. Always aim to keep perp widths to 10 mm if possible, since larger or smaller perps spoil the appearance of the face work.

It is good practice to set out a wall 'dry' before laying the bricks in mortar, especially if the wall has to be built to specific linear dimensions. These linear dimensions must be established by careful measurement, in accordance with the working drawings.

Setting out the first course dry means that perp joint sizes can be checked and either tightened or opened so that the wall will fit within the specified dimensions. This is often referred to as 'setting out the bond'.

A well-designed wall will have overall dimensions that work to fit full brick sizes. However, this is not always possible, and if varying the joint sizes does not make things fit, the bricklayer will have to employ other methods.

One method is to reverse the bond. Remember, this means that rather than having matching headers (or stretchers) at either end, the bonding arrangement is reversed so that the wall has a header at one end and a stretcher at the other end.

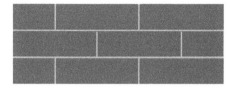

▲ Figure 4.22 Reverse bond

If this does not produce a wall with uniform joints within the overall length measurement, the only choice is to introduce cut bricks in each course. Placing cut bricks within a course of brickwork is called broken bond. The smallest cut allowed in the face of a half-brick-thick wall is known as a half bat (or batt) and measures 102.5 mm (the width dimension of the header face of a full brick).

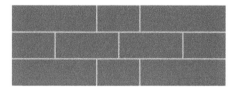

▲ Figure 4.23 Broken bond (note the placement of three-quarter bricks in relation to the half bat)

Cut bricks in broken bond should be placed as near to the centre of the wall as possible. When a half bat is built into a course of brickwork, the courses above and below it will contain two three-quarter cuts to maintain the bond.

ACTIVITY

In a workshop or on a smooth level floor, ask your tutor or trainer to mark wall length dimensions that are not whole brick sizes. Work with someone else to decide which method you will use to establish a good bonding arrangement.

The hand tools needed to cut bricks to specific dimensions were listed earlier in this chapter. Now we will look at the method of cutting bricks safely and accurately.

3.2 Cutting bricks

Remember, before starting any work task, consider the potential hazards. Make sure nearby workers are also aware of the risks. As well as using basic PPE, wear safety glasses or goggles to protect your eyes from flying brick fragments.

Cutting bricks using hand tools can only be mastered with experience, so be prepared to have disappointments at first. The main cutting tools are a club (or lump) hammer and a brick bolster.

Another useful tool not mentioned previously is a brick gauge, which is used to produce accurate cuts. Consistently producing half and three-quarter cut bricks is made simpler using this tool to accurately position the blade of the bolster chisel on the face of the brick being cut. A brick gauge can be made from timber available on site.

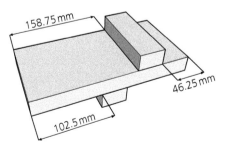

158.75 mm
46.25 mm
102.5 mm

▲ Figure 4.24 Brick gauge

Whether you use a brick gauge or a tape measure and pencil to mark the brick, always aim for accuracy. It will be easier to produce a finished wall to a high standard of appearance if the bricks have been cut accurately.

Many bricks are brittle and shatter easily. To improve the chance of cutting successfully, it is good practice to place the brick on a small mound of sand to act as a cushion. Alternatives such as sacking or old carpet can also be used. Make sure the whole surface area of whichever face is on the cushion is supported.

INDUSTRY TIP

Some bricklayers carry a small piece of old carpet in their tool kit specifically to use as a cushion when cutting bricks by hand.

Study the step-by-step guide to see how to cut bricks safely and accurately.

Step by step
Cutting bricks safely and accurately

Step 1 Mark the position of the cut on the face, the opposite side and the bed of the brick with a pencil.

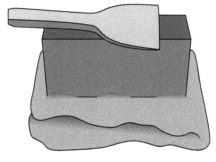

Step 2 Place the brick with the face uppermost on a mound of sand or other cushion.

Step 3 Place the blade of the bolster slightly on the waste side of the pencil mark, then strike the bolster lightly but firmly with the club hammer.

Step 4 Turn the brick over and repeat Step 3 on the opposite side of the brick.

Step 5 Turn the brick so that the bed is uppermost and strike the last blow. If the strength of the blow is adjusted correctly, this should complete the operation (remember – experience counts!). If the brick does not break as desired, repeat from Step 3 until a clean break is achieved.

HEALTH AND SAFETY

If you need to cut a large quantity of bricks to size, you will be able to work more quickly and with less waste if you use a powered masonry saw. Never use powered cutting equipment unless you have been properly trained and certified as competent.

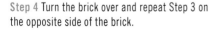

INDUSTRY TIP

Some bricks are hard and shatter easily. It can help to gently press the edge of the palm of the hand holding the bolster against the face of the brick as you are cutting. This will 'damp' the vibrations going through the brick and help to prevent it shattering.

▲ Figure 4.25 Damping vibrations using the edge of the palm of the hand

The aim is to produce clean, sharp edges, especially on the face side of the cut brick. This will contribute to a good appearance in the finished wall.

If there are rough projections or edges, a scutch hammer can be used to trim the cut brick and make precise adjustments. A brick hammer can also be used to trim bricks, but it is less precise and usually used to quickly produce rougher cuts where appearance is less important.

HEALTH AND SAFETY
Do not forget to use appropriate PPE when cutting bricks. As well as wearing a hard hat, safety boots and hi-vis clothing on site, gloves and eye protection are essential.

If a powered masonry saw is used, protect your hearing. Hearing damage can be permanent.

ACTIVITY

Try cutting six bricks:
- three bricks as half bats (102 mm)
- three bricks as three-quarter cuts (159 mm).

Ask someone else to check your cuts for accuracy and neatness.

3.3 Forming perp joints for brickwork

Rolling mortar to produce a bed joint for brickwork is essentially the same as for blockwork (see page 92 in Chapter 3). However, forming a perp joint for brickwork is quite different and requires continuous practice.

INDUSTRY TIP

You might hear experienced bricklayers referring to perp joints as 'cross joints'.

Preparing the mortar is important to produce perp joints quickly and consistently. Study the following step-by-step guide to see how to do this.

Step by step
Forming a perp joint on a brick

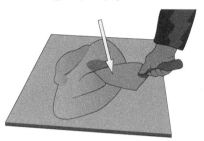

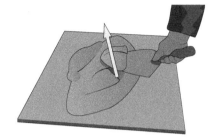

Step 1 With a mound of mortar on the spot board, tap the tip of the trowel into the mortar to produce a V-shaped groove.

Step 2 Insert the tip of the trowel into the mortar about 50 mm below the V-shaped groove and lift out a small amount of mortar.

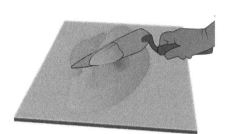

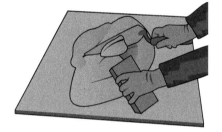

Step 3 Flip the mortar from the tip of the trowel towards the centre of the blade and shake the blade downwards firmly to spread the mortar a little. It will take practice to get this action right.

Note: When this action is done correctly, the mortar will stick to the blade of the trowel even when it is turned upside down.

Step 4 With the brick held almost vertically in one hand and the trowel in the other, spread (or 'butter') the mortar on the blade of the trowel over the header face of the brick.

There are a number of techniques to butter a brick, but the mortar should always cover the entire header face of the brick. The practice of 'wiping' a small amount of mortar on the front and back edges of the header face will not produce a weatherproof joint. This practice is sometimes called 'tipping and tailing' and is used by some bricklayers to speed up the rate of production. It can lead to problems later if moisture remaining in the gaps created in the perp joint freezes during cold weather. The perp joint must be full, without any gaps.

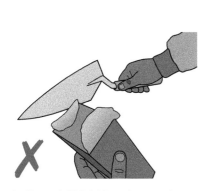

▲ Figure 4.26 Bricklayer incorrectly 'tipping and tailing' the perp joint

▲ Figure 4.27 Bricklayer correctly making a full perp joint

Developing the skills to correctly form mortar bed joints and perp joints takes a lot of practice. Accuracy and speed come from forming good habits and following effective routines of work.

3.4 Building quoins in brick

Having established the bond for the face of the wall, the usual method of working is to build quoins at each end of the wall. A string line is then set up between the quoins as a guide for building the infill.

Chapter 5 provides a lot of detail on how to build quoins for cavity walls, and the principles for building a quoin in a single leaf of brickwork are similar.

▲ Figure 4.28 Racking back brickwork

When a return quoin or corner is built, it is usually set out as a right angle (90°). If the quoin does not include a return, it is known as a 'stopped end'. In either case, the courses will be 'racked back', which means that the bricks form steps as the courses are laid.

Study the step-by-step guide to see the sequence of work when building a quoin in brick. The first courses in the quoin described extend four bricks from the corner in each direction.

Step by step
Building a quoin in brick

Step 1 Lay four bricks on a prepared mortar bed, aligned with the face line of the wall.

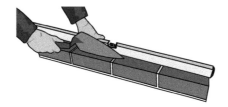

Step 2 Carefully level the four bricks, tapping them to match the bottom edge of the spirit level.

Step 3 Carefully line the faces of the bricks, tapping them to line up with the spirit level. Note that the spirit level is used simply as a straight edge.

➡

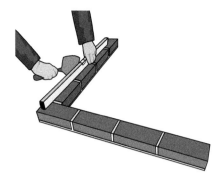

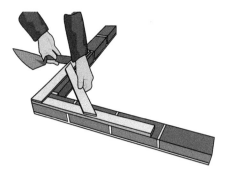

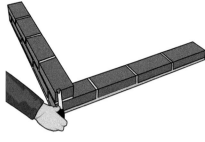

Step 4 Lay the four bricks in the return and repeat the levelling and lining-in steps.

Step 5 Use a builder's square to check the return is at right angles to the face and adjust as needed.

Step 6 Lay the second course to the face line and check the gauge is 75 mm for a brick and bed joint. Until you have more experience, check each course as the work proceeds. Using a gauge rod can make the job quicker.

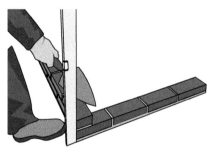

Step 7 Level the course away from the corner. Do not adjust the gauged corner brick. Make any adjustments needed to the other bricks in the course.

Step 8 Plumb both ends of the course and line-in between the two plumb points.

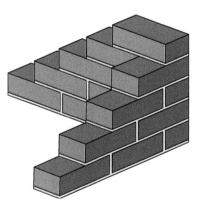

Step 9 Repeat the sequence for each course until the quoin is completed.

Step 10 Finally, range the work. This means using the spirit level as a straight edge but this time presenting it diagonally along the stepped courses to align the ends of each course.

Lining-in each course and ranging the quoin will make it easier to produce an accurate face plane when building the brick infill between the quoins.

Remember, you will be able to work more quickly if you practise good habits. Follow this routine:

1 Gauge the course.
2 Level the course.
3 Plumb each end of the course.
4 Line-in the face of each course.

3.5 Laying to the line

A string line can be attached to the finished quoins at both ends of the wall to act as a guide when building the infill brickwork between the corners. Laying to the line is a skill that needs extensive practice in order to build speed and accuracy.

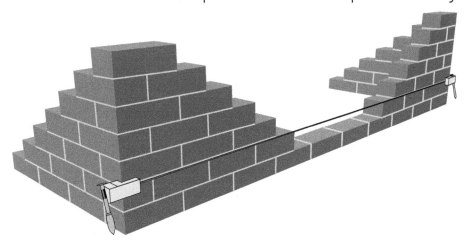

▲ Figure 4.29 String line attached to quoins

Study the step-by-step guide to see how this should be done.

Step by step
Laying to the line

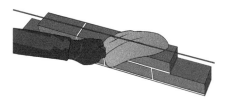 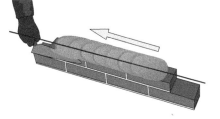 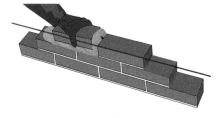

Step 1 Pick up enough rolled mortar for two or three bricks. Prepare the mortar by rolling it on the spot board. Avoid picking up too much, as it will become difficult to manage.

Step 2 'Throw' the mortar into a line along the position where the bricks are to be laid and adjust as needed to produce the bed joint. (See Chapter 3 for more on how to produce a bed joint.)

Step 3 Lay a brick, making sure the starting position is higher than the string line along the entire length of the brick.

➡

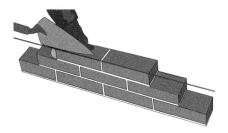

Step 4 Gently tap the brick so that the top arris lines up with the string line. Note the position of the trowel when tapping the brick. Make sure the brick does not touch the line. If it does, it will push the line out of accurate alignment and the wall will not be straight.

Step 5 Make sure the bottom arris lines up with the course below, and check the perps are plumb with the perps below.

Remember the key points when laying bricks to the line:

- Make sure the top arris of the brick is level with the line along its full length.
- Never lay the brick touching the line.

INDUSTRY TIP

When laying to the line, you must take care to ensure the top and bottom arris of each brick is accurately aligned. Otherwise, shadows can be created across the face of the wall which spoil the appearance. This is often referred to as 'hatching and grinning'.

It is especially important when constructing face brickwork to keep the perp joints plumb throughout the height of the wall. 'Eye' down the face of the wall to align the perps in each course with those in alternate courses below.

▲ Figure 4.30 Make sure the perp joints are plumb and line up – not like this!

▲ Figure 4.31 Laying to the line

3.6 Building junctions in brick

There are many points in structures where one wall forms a junction with another. A junction is formed by bonding the masonry in a specified way to create a stable and strong feature. In a half-brick-thick wall, the bonding arrangement is not unlike that used in broken bond, as discussed earlier.

Study Figure 4.32 to see how the header face of a junction wall is bonded into the main wall using three-quarter cut bricks.

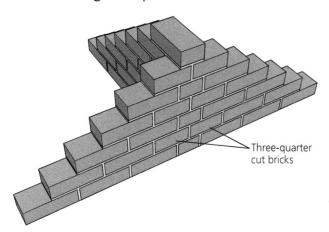

Three-quarter cut bricks

▲ Figure 4.32 Junction in a half-brick-thick wall (note the use of three-quarter cut bricks to form the bond)

There are other ways that masonry can be bonded for stability. The method chosen will depend on factors such as access requirements and the planned sequence of work. One method uses **indents** in the main wall.

The indents are the width of the wall that will be built later, with allowance for a mortar joint either side to make sure the junction is solid and stable.

> **KEY TERM**
>
> **Indents:** holes or pockets accurately formed at each course or block of courses in the main wall as building proceeds

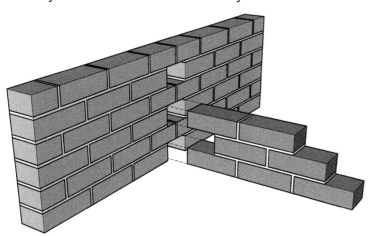

▲ Figure 4.33 Indents formed ready to receive a junction wall later

---

INDUSTRY TIP

To make it easier to build a junction wall into the indents at a later stage, form them slightly wider than a 10 mm joint either side.

As mentioned in Chapter 3, a simpler way of providing for a junction wall at a later stage is to use reinforcing mesh. This is carefully built into the bed joints on the vertical line of the wall that will be added later. The mesh is built in as work on the main wall proceeds, in accordance with the specifications.

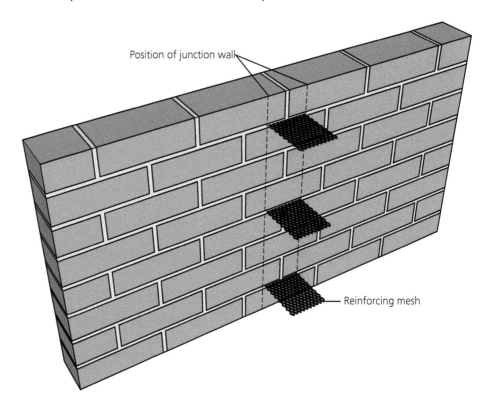

Position of junction wall

Reinforcing mesh

▲ Figure 4.34 Reinforcing mesh built into bed joints to receive a brick junction wall later

3.7 Forming joint finishes

A wall built in facing brick will need a joint finish to be formed for the bed and perp joints. It takes patience and practice to form joint finishes that are visually acceptable in appearance and weatherproof.

The tools used to form a joint finish are sometimes referred to as jointing irons. Ironing the joint will smooth and compress the surface of the mortar to seal it against the entry of moisture.

▲ Figure 4.35 Ironing the mortar joints to seal them

The most commonly used joint finish is half-round, which is sometimes referred to as concave. When producing this finish, joint all the perp joints first, before working on the bed joints. This reduces the number of untidy projections which can form where the perp and bed joints meet.

Table 4.4 shows some different joint finishes that can be used for a brick wall.

▼ Table 4.4 Joint finishes for a brick wall

Type of joint finish	Description
Half-round joint	Often referred to as concave or 'bucket handle', this is probably the most commonly used joint finish. It can be produced quite quickly and disguise irregularities in the shape of the brick. The action of ironing (or tooling) the joint presses the mortar against the arris of each brick, giving greater resistance to the effects of poor weather.
Recessed joint	This type of joint finish is formed using a purpose-made tool to remove mortar from the joints to a specified depth. The tool can be a simple timber block cut to the right shape on site, or it can be a special tool with wheels, often referred to as a chariot.

➡

▼ Table 4.4 Joint finishes for a brick wall (continued)

Type of joint finish	Description
Weather struck joint	This type of joint finish is formed by shaping the joint with a trowel (preferably a pointing trowel) so that it slopes to allow rainwater to run off it. The top of the joint is set back from the face slightly and the bottom of the joint remains flush with the brick below. The perps are also angled, with the left side of the joint sloped in slightly.
Flush joint	This type of joint finish is often used where a specific appearance is required. It is produced by smoothing and compacting the mortar with a hardwood timber block as the work progresses. The disadvantage is that it is difficult to achieve a weather-tight finish without giving the joint the appearance of being wider than it actually is.

Always keep in mind that it is the bricklayer's responsibility to ensure that the completed work meets the specification and has a good appearance without defects. Take time to get the jointing right, so that the brickwork is sealed against the effects of bad weather.

The quality of the work produced will affect how long the wall lasts and performs its intended function.

▲ Figure 4.36 Take the time to produce a joint finish to a high standard

▲ Figure 4.37 Poorly jointed work at the corners of a brick quoin

Frequently check your work for gauge, level, plumb and line when building quoins and junctions, and develop your skills to lay to the line quickly and accurately. Taking care will produce masonry that not only looks good but will also stand the test of time.

Test your knowledge

1 Which type of drawing gives a view showing each face or side of a work task?

   a Elevation

   b Section

   c Plan

   d Detail

2 How many bricks are there in 1 m² of a half-brick-thick wall?

   a 10

   b 20

   c 40

   d 60

3 Which tool is used to precisely trim and shape cut bricks?

   a Bolster chisel

   b Scutch hammer

   c Club hammer

   d Jointing chisel

4 What is the main reason for using a brick bond where each course overlaps the course below?

   a To save materials

   b To reduce weight

   c To spread loadings evenly

   d To improve the appearance

5 What is the width in millimetres of a standard brick?

   a 100.5

   b 102.5

   c 105

   d 120.5

6 What is the allowable tolerance that can be applied to a 10 mm perp joint?

   a ±2 mm

   b ±3 mm

   c ±4 mm

   d ±5 mm

7 Why is setting out 'dry' used?

   a To establish the face plane

   b To establish the ranging

   c To establish the correct gauge

   d To establish the correct bond

8 Which bonding arrangement can be used when the overall wall measurement will not work to exact brick sizes?

a Filled

b Return

c Stopped

d Reverse

9 If a brick wall must be set out to include cut bricks in each course, what is the bonding arrangement called?

a Damaged

b Broken

c Halved

d Trimmed

10 To what does the term 'indents' refer when building a brick wall?

a An apprentice's qualification papers

b The dips in an uneven concrete foundation

c Pockets left in a wall to form a junction

d Marks on the surface of a bolster chisel

BUILDING CAVITY WALLS

INTRODUCTION

This chapter considers:
- the reason for using cavity walls
- the range of materials and essential tools and equipment used by the bricklayer when building cavity walls
- how cavity walls function in masonry structures.

Building cavity walls requires construction methods that are more technically demanding than those used when building a solid wall in brick or block. High standards of work are essential in the construction process to make sure the finished cavity wall does its job.

LEARNING OUTCOMES

After reading this chapter, you should:
1. understand the reasons why cavity walls are used
2. know how to select the correct tools and equipment for cavity wall construction
3. know how to identify and select the correct materials for cavity wall construction
4. know the correct methods used for cavity wall construction.

1 WHY CAVITY WALLS ARE USED

In the past, brick buildings were often constructed with solid walls. Beginning in the 1920s, and especially from the 1950s onwards, cavity walls began to be used because of their greater effectiveness in preventing moisture from entering the living and working spaces of buildings. The design of a cavity wall is two leaves (or skins) of masonry with a gap (cavity) between them.

▲ Figure 5.1 Older brick houses were built with solid walls

1.1 Preventing damp

The cavity between the two leaves of masonry forms a barrier, across which moisture will not travel. By contrast, when a solid wall becomes wet it allows moisture to travel more easily to the inside of the wall, creating a damp environment inside the building.

Preventing damp in a building is important for two reasons:

- Damp affects the health of the occupants.
- Damp causes the **fabric** of a building to deteriorate.

▲ Figure 5.2 Damp in a building

1.2 Other reasons

A cavity wall is usually wider at its base than a solid wall, so it places lighter loads on the foundation since the load is spread across a wider area. This means that the design of the foundation for cavity walls can be simplified or reduced in thickness, since it does not need to support the greater concentrated weight of a solid wall.

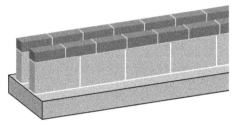

▲ Figure 5.3 A cavity wall is wider at the base

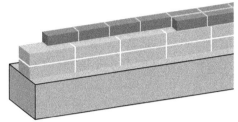

▲ Figure 5.4 A narrower solid wall concentrates loading on the strip foundation

ACTIVITY

Review the information about foundation design in Chapter 1 and note the different thicknesses of concrete strip foundations.

A while after their introduction, it was realised that cavity walls also give building designers the opportunity to install several different types of wall insulation. (You will learn about insulation later in this chapter and much more at Level 2.)

Preventing heat transfer in buildings is an increasingly important aspect of building design as the drive to reduce **carbon emissions** continues.

Early cavity-wall designs allowed ventilation of the cavity to occur, with the idea of keeping it dry. This was achieved by means of specially designed components called air bricks, built into the masonry. Later, a sleeve or **duct** was added to ventilate the interior of the building instead of ventilating the cavity.

KEY TERMS

Carbon emissions: the release of carbon into the atmosphere (also referred to as greenhouse gas emissions) contributing to climate change

Duct: a tube or channel which allows the passage of a liquid or gas

▲ Figure 5.5 A ducted air brick

To meet modern requirements to control carbon emissions, cavity walls and other elements of a structure are now designed to make a building airtight on completion. By controlling the movement of air, while at the same time providing precisely controlled ventilation, a building can be made more energy efficient.

INDUSTRY TIP

Ventilation components are sometimes installed by a bricklayer when suspended floors are part of the building design. These are often referred to as telescopic vents, because their length can be extended or reduced by sliding the inner section over the outer section. They allow air from above finished ground level to be channelled under the suspended floor.

---

2 SELECTING THE CORRECT TOOLS AND EQUIPMENT FOR CAVITY WALL CONSTRUCTION

As mentioned in previous chapters, it is always a good idea to make a list of the tools and equipment that will be required for a job. Doing this before you start work makes things run as smoothly as possible. Remember, the tools needed for any masonry task can be split into three main groups:

- laying and finishing
- checking
- cutting.

INDUSTRY TIP

Storing tools in a secure place is important. If other workers have access to your tools, they may not look after them as carefully as you do and they could become unsafe to use.

2.1 Tools

Table 5.1 shows the main tools you will use for masonry tasks. There is more information about using cutting tools to cut bricks and blocks by hand in Chapters 3 and 4.

▼ Table 5.1 Tools for masonry tasks

Tools for laying and finishing	
Trowel 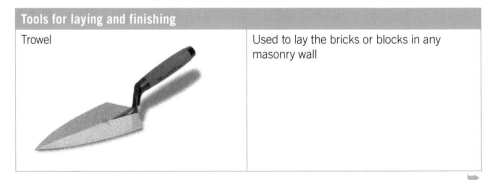	Used to lay the bricks or blocks in any masonry wall

▼ Table 5.1 Tools for masonry tasks (continued)

Tools for laying and finishing	
Pointing trowel	Used for shaping mortar joints when producing face brickwork
Jointer	Used for forming a joint finish to bed and perp joints in face work

Tools for checking	
Tape measure	Used to set out and check dimensions
Spirit level	Used to make sure the work is level and plumb Spirit levels are available in a range of sizes, including a version that can fit in a pocket
Line and pins	Used as a guide, to check the bricks or blocks are aligned accurately
Corner block	Used to position a line and pin on corner (or quoin) blocks

Corner block

▼ Table 5.1 Tools for masonry tasks (continued)

Tools for cutting	
Club hammer	Used with a brick bolster to cut bricks or blocks accurately Also called a lump hammer
Brick bolster	Used like a chisel with a club hammer to cut bricks or blocks accurately to the required dimensions
Brick hammer	Used to cut and trim bricks or blocks quickly, where accuracy is not critical
Scutch hammer	Used to accurately and finely trim and shape bricks and blocks

ACTIVITY

Make a list of all the tools described so far in this chapter. Go online and find out the best price for each item. Add up the prices of the listed tools to see how much your tool kit could cost.

2.2 Equipment

Table 5.2 shows some of the equipment needed to build cavity walls.

▼ Table 5.2 Equipment required to build cavity walls

Equipment	Description/How it is used
Spot boards	Boards that have mortar placed on them within easy reach of the bricklayer along the length of a wall
Shovel	Used to move mortar and loose materials, or for mixing mortar by hand
Wheelbarrow	Essential for moving materials to the work location
Brick tongs	A specialist piece of equipment used for efficiently and safely picking up a number of bricks at once

▼ Table 5.2 Equipment required to build cavity walls (continued)

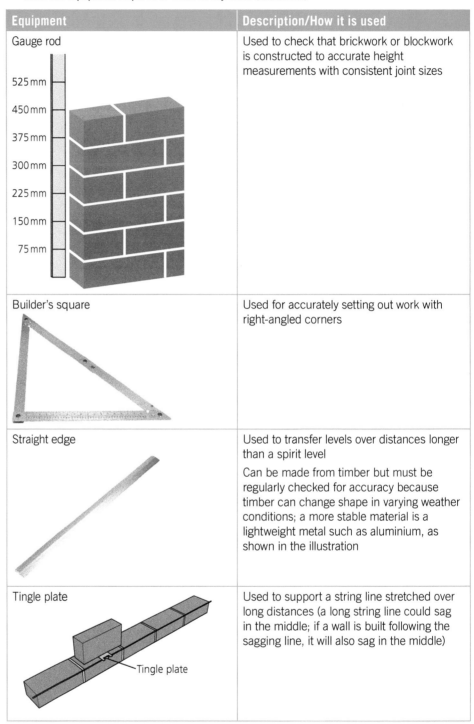

Equipment	Description/How it is used
Gauge rod 525 mm 450 mm 375 mm 300 mm 225 mm 150 mm 75 mm	Used to check that brickwork or blockwork is constructed to accurate height measurements with consistent joint sizes
Builder's square	Used for accurately setting out work with right-angled corners
Straight edge	Used to transfer levels over distances longer than a spirit level Can be made from timber but must be regularly checked for accuracy because timber can change shape in varying weather conditions; a more stable material is a lightweight metal such as aluminium, as shown in the illustration
Tingle plate Tingle plate	Used to support a string line stretched over long distances (a long string line could sag in the middle; if a wall is built following the sagging line, it will also sag in the middle)

INDUSTRY TIP

A bricklayer can make a gauge rod on site from available lengths of timber. The gauge markings are made using saw cuts at the correct positions throughout the height of the rod.

A type of gauge rod called a 'storey rod' is used to check the accuracy of height measurements at specific points of a building, such as the tops of windows or upper-floor levels.

③ IDENTIFYING AND SELECTING THE CORRECT MATERIALS FOR CAVITY WALL CONSTRUCTION

As mentioned in previous chapters, it is good practice to make a list of the types and quantities of materials required for a job and to ensure they are available before starting work.

This reduces the likelihood of running out of materials mid-project, which could cause delays while additional materials are brought to the work location. On the other hand, if there are more materials at the work location than required, the bricklayer will waste time moving them after the work is completed.

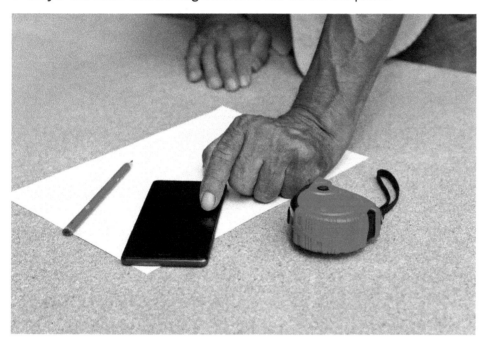

▲ Figure 5.6 Calculate what materials you need before starting work

3.1 Calculating quantities

Methods of calculating the required quantities of materials are discussed in Chapter 1.

Remember that calculating quantities of masonry components is simplified by keeping in mind the number of bricks and blocks in 1 m² of a half-brick-thick wall:

- There are 60 bricks to 1 m² of wall.
- There are 10 blocks to 1 m² of wall.

By remembering these two figures, you simply need to work out the area of the face of the wall in square metres and multiply the result by 10 or by 60, depending on which material quantity is being calculated.

> **INDUSTRY TIP**
>
> Area is calculated in m² by multiplying the height of a wall by its length. By including an amount for waste (say 5%), we can make sure we do not run out of materials and avoid wasting time.

3.2 Checking suitability

When preparing for masonry construction tasks, the first step is to get familiar with exactly what is being built.

The main sources of information are the specification and working drawing, which will help the bricklayer to understand all the details of the job. These documents will tell you what types of materials are to be used. If the wrong bricks are used, they may not be suitable for the location and conditions in which the cavity wall is being built. The wrong blocks may have inadequate strength or insulation properties.

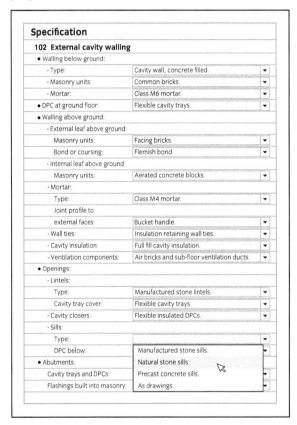

▲ Figure 5.7 Specification

The bricklayer should also check the condition of the materials before using them. If damaged bricks or blocks are used, this can affect the appearance of finished work as well as reducing the strength of the cavity wall.

INDUSTRY TIP

Sometimes, a large proportion of a batch of materials delivered to site might be damaged. Alert your supervisor as soon as possible, so that unsuitable materials can be removed from use.

Carefully checking the suitability of materials and resources before starting construction maintains high standards of work. Expensive alterations might be required if unsuitable materials or work methods are used.

The main materials used in the construction of cavity walls are, of course, bricks and blocks, the characteristics of which are discussed in detail in Chapters 3 and 4. However, since it is important to understand the characteristics of these materials in relation to cavity wall construction, we will review them here.

3.3 Bricks and their characteristics

Bricks are delivered to site in packs, which are usually off-loaded mechanically by crane or forklift for efficiency and to reduce the risk of damage to the materials.

It is important to choose appropriate bricks for a particular location or type of structure. The architect and their design team are responsible for specifying the right brick for the job and will consider the requirements for cavity walls when making the choice.

Bricks fall into three main categories:

- facing bricks
- engineering bricks
- common bricks.

Facing bricks

Facing bricks can form the 'face' of the building and are available in an extensive range of colours and textures. They are usually manufactured from clay, which is easily moulded. They can be produced in solid form, with holes in them (known as 'perforated') or with a depression in the top surface known as a 'frog'.

▲ Figure 5.8 Facing bricks

The moulded soft clay is converted into durable bricks by a process known as 'firing'. This involves heating the bricks in a **kiln** to temperatures between 900°C and 1250°C. The process may produce bricks that vary a lot in size since the heating process causes the materials to change their form, resulting in shrinkage and distortion.

▲ Figure 5.9 Brick kiln

Facing bricks can also be manufactured from concrete or sand and lime.

Sand/lime bricks are moulded to shape under high pressure in hydraulic presses. They are then subjected to high-pressure steam in a large piece of equipment called an autoclave to **cure** the bricks. Because they readily absorb moisture, they should not be used in areas exposed to wet conditions.

This process produces bricks that have very little variation in overall dimensions, making it easier to produce uniform and consistent work.

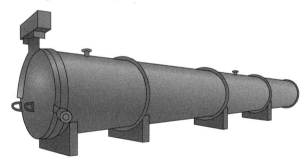

▲ Figure 5.10 Brick autoclave

INDUSTRY TIP

Sand/lime bricks are sometimes referred to as calcium silicate bricks, as they are composed of a fine aggregate which is bonded together by **hydrated** calcium silicate. Pigment can be added to produce bricks in a range of colours.

▲ Figure 5.11 A sand/lime brick wall

Concrete bricks are manufactured from a mix of aggregate, cement and water. They can be coloured during manufacture and given a range of textures if desired. Since they are heavier than clay bricks, concrete bricks are effective in reducing noise transmission and offer good fire protection.

▲ Figure 5.12 A concrete brick wall

Engineering bricks

Engineering bricks have high compressive strength. This means that they can resist squeezing forces created by the weight of a high-rise structure, like a block of flats, or the squeezing forces in a structure that carries a lot of weight, like a bridge.

▲ Figure 5.13 A railway bridge built with engineering bricks

Also, because of the type of clay they are made from and the temperature at which they are fired, engineering bricks do not absorb water. This makes them suitable for use in walls below ground level.

Common bricks

Common bricks are lower quality bricks that are usually used in locations where the finished work will not be on show. For example, they could be used in constructing internal partition walls that adjoin a cavity wall.

▲ Figure 5.14 A common brick with frog

ACTIVITY

Carry out some research and list two situations in which a brick wall would be built below ground level.

IMPROVE YOUR ENGLISH

After reading the descriptions of the different types of brick:

1 Discuss with a partner which type of brick would be suitable for use in the basement of a three-storey house.
2 Write down a description of your chosen brick in your own words and give two reasons for your choice.

3.4 Blocks and their characteristics

Blocks used in the construction of cavity walls fall into two main categories: lightweight insulation blocks and dense concrete blocks.

Lightweight insulation blocks

Lightweight insulation blocks are usually specified for the internal leaf of a cavity wall. They are made from materials that reduce heat transfer through the wall, making it more energy efficient.

Because they are lightweight, they are easy to work with and can be readily cut and shaped. However, especially when they are dry, they generate a lot of fine dust when being cut to shape or moved from storage.

▲ Figure 5.15 Lightweight insulation blocks

INDUSTRY TIP

To reduce the dust generated by cutting or moving lightweight insulation blocks, they can be lightly sprayed with water. However, be careful not to saturate them.

HEALTH AND SAFETY

It is essential to use suitable PPE to protect yourself against the effects of fine dust. Not only can it irritate your eyes, but breathing it in can be damaging to your lungs.

Dense concrete blocks

Dense concrete blocks can be specified in cavity walls where the outer leaf is not built in brick. (To provide a suitable finish, a coating of sand/cement mix called render or another type of finish will need to be used.) They are also used extensively in work below ground, such as in the footings of a cavity wall.

3.5 Characteristics of other materials and components

Cavity walls can be built using other materials, such as stone, timber and concrete. Different construction methods are used for these materials, some of which are discussed in Chapter 1.

Mortar

Bricks and blocks are bonded together using mortar to create the joints. Care must be taken when mixing mortar for use in cavity walls, to produce a material that is easy to use ('workable') and long lasting when set.

A specification for a cavity wall may give precise details about:

- the colour of the mortar
- the proportions of the different materials in the mortar mix
- any additives such as a plasticiser that are required.

INDUSTRY TIP

When using additives in a mortar mix, always check the manufacturer's instructions carefully. Using the wrong quantities or the wrong type of additive could result in a mix that does not meet the specification.

▲ Figure 5.16 Mortar on a spot board

As discussed in Chapter 3, mortar for masonry construction is produced by mixing cement, sand and water to a workable consistency. Remember that mixing the correct proportions of each material is accomplished by using a specified ratio.

ACTIVITY

Review the information in Chapter 3 about using clean water when mixing mortar. Write down the possible consequences of not using water of suitable quality.

Wall ties

Different types and sizes of wall tie are available to suit various cavity widths.

Ties were previously manufactured from a type of steel which suffered from long-term failure due to corrosion and rust. Modern ties are commonly made from stainless steel, which is much more durable. More rarely, wall ties have been manufactured using polypropylene (a type of plastic).

The design of wall ties includes a feature known as a 'drip'. This is important in preventing moisture from crossing the cavity along the tie. There is more about this later in the chapter.

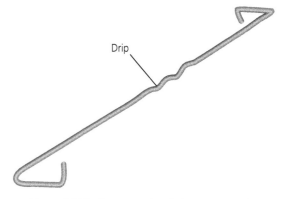

Drip

▲ Figure 5.17 Stainless-steel wall tie

Damp-proof course

A damp-proof course (DPC) is an essential component of cavity wall construction and will be specified in several locations as part of the design. In this chapter we will discuss horizontal DPCs, which are installed to prevent **rising damp**. A process called capillary attraction draws moisture from the ground into tiny gaps in the mortar bed or masonry materials, acting like a sponge, and this moisture travels upwards in the wall. A horizontal DPC breaks the path of the moisture.

DPCs can also be used vertically around openings and in the form of a continuous tray above openings.

In a cavity wall, a flexible DPC manufactured from polythene is most commonly used. There are other types of DPC, such as pitch polymer and bitumen felt, but these are used less frequently.

KEY TERM

Rising damp: when moisture from the ground travels up through the walls by capillary action

▲ Figure 5.18 A roll of polythene DPC

INDUSTRY TIP

If a DPC such as bitumen felt is specified, be careful when using it in cold weather. In low temperatures bitumen felt becomes stiff and brittle and can crack easily. If a DPC has cracks, it will not prevent damp entering a building.

Insulation

As mentioned previously, insulation can be installed in a cavity wall to reduce heat transfer. Insulation materials are relatively fragile and easily damaged, so the bricklayer should handle them with care. Damaged materials may not function as intended and the efficiency of the insulation could be reduced.

Insulation can be installed in a cavity wall in three ways:

- full fill
- partial fill
- injected.

Methods of installing insulation and details about the materials it is made from will be covered in more detail at Level 2.

4 METHODS OF CONSTRUCTING CAVITY WALLS

After carefully identifying, checking and preparing the resources, the next stage is to set out the work area and position the materials and components ready for building a cavity wall.

Always give careful consideration to health and safety matters, to protect yourself and others working nearby when moving and handling materials and components. Make sure the correct PPE is used and that it is in good condition. Chapter 6 discusses health and safety matters related to bricklaying work in more detail.

▲ Figure 5.19 A bricklayer wearing PPE

4.1 Positioning materials in the work area

Always move and handle bricks and blocks with care. Bricks can be sharp and many have a rough textured surface. Masonry materials are heavy and bulky, making them awkward to handle. This is especially the case with dense concrete blocks. Remember, it is easy to injure your hands and fingers by trapping them between bricks and blocks when stacking them.

A number of methods can be used when stacking bricks ready for laying. For example, stacks of bricks can be positioned on edge, with two rows of six bricks in each layer and up to 12 layers in each stack. See Figure 5.20.

Stacks can be higher, but when in doubt discuss this with your supervisor or a more experienced bricklayer.

> ### INDUSTRY TIP
>
> Stacking bricks on edge allows the use of brick tongs to move six bricks at a time. This can speed up the preparation process.

Another method is to carefully stack the bricks flat in neat groups, making sure the arrangement is stable. Whichever method is used, it may be necessary to spend some time levelling the ground to make sure the materials are positioned safely.

> **HEALTH AND SAFETY**
>
> Rather than moving heavy materials by hand, arrange to use mechanical handling methods such as a forklift or crane wherever possible. The operators of these machines are trained in safe working practices, so always cooperate with their directions and be alert.

▲ Figure 5.20 Bricks stacked properly (note the use of the proper tool for moving bricks)

▲ Figure 5.21 Bricks stacked carefully

> ### INDUSTRY TIP
>
> When arranging bricks in stacks ready for work, select bricks randomly from a minimum of three packs to mix the colour variations. This will avoid creating obvious bands of colour (called 'banding') in the cavity wall face.

ACTIVITY

As you travel around, try to identify examples of banding in local brickwork. If you have a phone with a camera, take pictures and discuss them with your tutor or trainer.

HEALTH AND SAFETY

When stacking bricks or blocks on a scaffold or raised working platform, be aware of the potential for overloading the access equipment. Always place stacks of materials where the working platform is supported, not where it spans between supports. Never stack materials above the height of the handrail.

For more on the Work at Height Regulations 2005, see Chapter 6.

Stacks of blocks are best arranged flat rather than on edge, six to eight layers high, on a firm and level base. Dense concrete blocks are best moved one at a time if moved by hand.

The distance that materials are placed from the wall is important for efficient working. Stacks of materials and spot boards for mortar should be placed around 600 mm from the face line of the wall, but some bricklayers prefer to increase the working space between the materials and the wall to around 900 mm if the work area is suitable.

When building a cavity wall, often three materials must be positioned: bricks, blocks and mortar. You could position a stack of bricks next to a stack of blocks next to a spot board of mortar and repeat this pattern along the length of the wall.

ACTIVITY

Draw a plan view sketch to show how stacks of materials are positioned in relation to a cavity wall to allow for efficient working practices. Use a ruler for neatness.

▲ Figure 5.22 Bricks, blocks and mortar ready for use to build a cavity wall on site

IMPROVE YOUR MATHS

Ask someone else in your learner group to make up some realistic dimensions for a cavity wall (length and height). Practise calculating the number of bricks and blocks needed for the imaginary wall and ask the other learner to check your calculations.

4.2 Positioning components in the work area

Lighter components such as wall ties should be positioned in small bundles along the length of the wall, ready for use. They could be placed in the space under raised spot boards, where they are easy to grasp but will not be stood on and damaged or pressed into soft ground and lost.

Insulation materials are light but can be bulky. Position them in a convenient location near the wall being built and make sure that packs or loose sheets of material are weighted down to prevent them being picked up by the wind and damaged.

INDUSTRY TIP

A simple way of preventing insulation sheets or packs from being blown away is to carefully place a dense concrete block or two on top of them.

4.3 Setting out the cavity wall

As mentioned previously, cavity walls consist of two leaves of masonry, most frequently face brick for the outer leaf and block for the inner leaf. In the following final sections of this chapter, we will focus on the cavity wall construction as part of the superstructure. (Keep in mind that when 'setting out' is referred to in these sections, it relates to cavity wall construction above the horizontal DPC, not the setting out of the building shape and position.)

The width of the cavity will be set when the footing masonry is constructed. This dimension will vary according to the design of the structure and whether or not insulation is to be built into the cavity. In the past, the width of the cavity was usually 50–75 mm. With the more frequent use of insulation in the cavity, the dimension is often increased to a width of 100 mm or more.

4.4 Installing the DPC

A horizontal DPC must be installed in both leaves of the cavity wall at finished floor level (FFL). This is at a minimum height of 150 mm above finished ground or path level around the building.

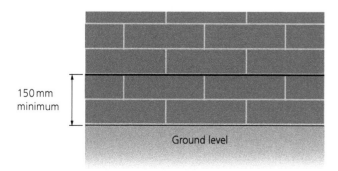

150 mm minimum

Ground level

▲ Figure 5.23 Minimum height of the DPC above ground level

The bricklayer is responsible for the correct installation of a DPC in a cavity wall and must give careful attention to the following points of good practice:

▲ Figure 5.24 Flexible DPC materials in stacks

- Store rolls of DPC correctly. Stack rolls on end, no more than three rolls high. Rolls of DPC that are stored on their side are prone to distortion, which can make them difficult to install. If possible, store at an even temperature.
- Make sure the specified material is used. Once the DPC has been built into the cavity wall, it will be very difficult to change it if it is not the correct type.

ACTIVITY

Go online and find a website for a builder's merchant. Search for DPC and list the different widths available.

Step by step
Installing DPC correctly

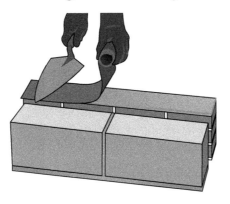

Step 1 Lay flexible DPC on a thin bed of mortar to protect it from possible puncture by hardened mortar projections in the previously laid work.

Step 2 If sections of flexible DPC have to be overlapped along the length of a wall, make sure the lap is a minimum of 100 mm. (If a DPC wider than 100 mm is specified, then the lap should be the same as the width of DPC used.)

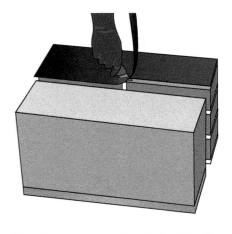

Step 3 Never allow the DPC to project into the cavity.

Step 4 If necessary, cut off projecting DPC with a craft knife.

DPC projecting into the cavity can form a platform for a build-up of mortar droppings. This can create a path for moisture to 'bridge' the cavity and enter the living or working area of the building.

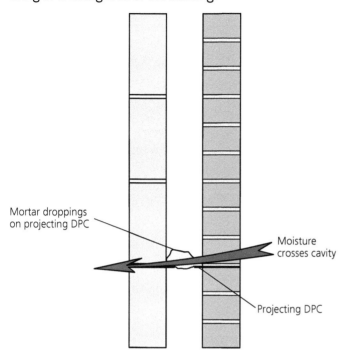

Mortar droppings on projecting DPC

Moisture crosses cavity

Projecting DPC

▲ Figure 5.25 Mortar droppings on projecting DPC allowing moisture to cross the cavity

Once the horizontal DPC is in place, the brick bond of the wall can be established.

INDUSTRY TIP

To keep the horizontal DPC in position until the wall is built above it, place individual bricks along its length at 1 m intervals to weigh it down.

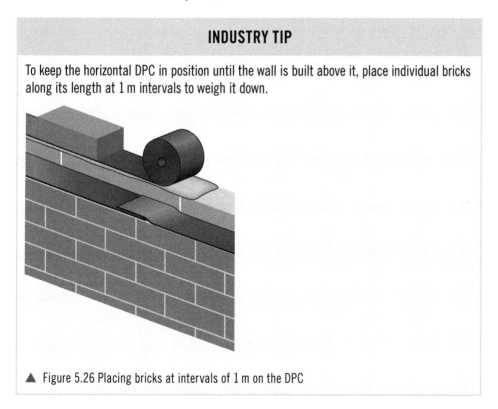

▲ Figure 5.26 Placing bricks at intervals of 1 m on the DPC

4.5 Establishing the bond

The most common bond used in cavity walls is stretcher bond. This is often referred to as half bond, because alternate courses of brick are laid with an overlap of approximately half the length of a standard brick. Block work is half bonded in a similar manner, with an overlap between courses of approximately half a block length.

▲ Figure 5.27 Stretcher (or half) bond

When the outer leaf of a cavity wall is constructed in face brick, correctly setting out and bonding the brickwork is vital to produce a wall that is both visually pleasing and structurally sound.

Take some time to establish the brick bond for the face of the cavity wall, setting out the first course of face brickwork 'dry'. Remember, this means positioning bricks without mortar to get the spacing right. Using this method will help you to decide how to make things fit within the design dimensions of the cavity wall.

Keep perps to 10 mm if possible. Although there is an allowable tolerance of plus or minus 3 mm, large perps spoil the appearance of the face work.

It may be necessary to cut bricks and blocks to create a suitable bond along the length of the wall and to fit within the design dimensions. The tools needed to cut masonry materials were described on page 141 and the methods of using these tools safely and efficiently are detailed in Chapters 3 and 4.

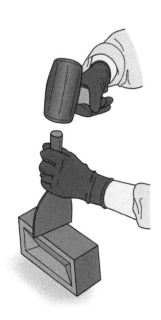

▲ Figure 5.28 A club hammer and brick bolster are used to cut bricks

Having established the bond for the face of the cavity wall, the usual method of working is to build quoins at each end of the wall. A string line is then set up between the quoins as a guide for building the infill.

4.6 Building quoins for cavity walls

Quoins are usually set out as a right angle (90°). Always lay and level bricks and blocks away from the corner point to maintain accuracy and efficiency, and follow a sequence of checks:

1 Gauge the course.
2 Level the course.
3 Plumb each end of the course.
4 Line-in the face of each course.

Gauge the course

Check each course for gauge as the work progresses. When you have more experience, you will be able to gauge after several courses have been laid and still keep your work accurate. Remember, standard gauge for one brick and a 10 mm joint is 75 mm.

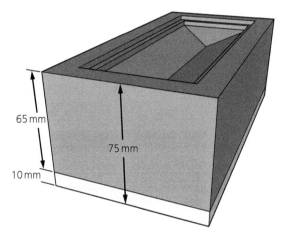

65 mm

75 mm

10 mm

▲ Figure 5.29 A standard brick and a 10 mm joint equal 75 mm

IMPROVE YOUR MATHS

Work out in millimetres the standard gauge for eight courses of brick. Ask your tutor or trainer to give you a few other examples to work out.

The gauge in superstructure masonry is measured from the horizontal DPC height, which usually corresponds to finished floor level (FFL). Height references such as FFL are called datum points and these points are established when setting out the building. Chapter 2 covers datums in more detail.

IMPROVE YOUR ENGLISH

Search online for two definitions of 'datum level'. Copy the definitions and compare your definitions with those of another learner in your group.

Gauge for each course of blockwork is 225 mm (a standard block 215 mm high plus a 10 mm joint). This means that one block course and a bed joint will equal the height of three brick courses and their bed joints throughout the height of a cavity wall.

Level the course

When levelling each course in a quoin, place the spirit level on top of the course of bricks to be checked and gently tap the bricks to line them up and get them level. If it seems necessary to keep tapping or striking the bricks and they do not move, it may be better to remove some bricks and adjust the thickness of the bed joint.

▲ Figure 5.30 Spirit level being used to level the top course of brickwork

Think of the brick or block at the corner point as a reference to level from once it has been gauged accurately.

Plumb the course

When plumbing each course, keep the spirit level stable by placing your foot against the bottom of it while holding the top with your free hand. Carefully adjust the bricks at either end of the course and look down the face of the wall to make sure it lines up with the spirit level.

INDUSTRY TIP

When levelling or plumbing brick courses in the quoin, remember that the spirit level is a precision instrument. Never strike the spirit level to align bricks.

Line-in the course

Lining-in the face of bricks between the plumbed ends of each course will help to produce a wall that has an accurate face plane. Rest the edge of the spirit level against the face of the bricks in each course without referring to any of the bubbles in the level – just use the level as a straight edge. Gently tap the bricks into line.

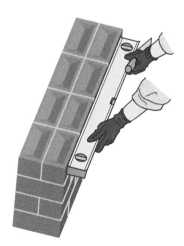

▲ Figure 5.31 Lining-in the course

Continue following this sequence for each course of the quoin to produce a corner that steps back (or 'racks back') to the top.

▲ Figure 5.32 Quoin

The final step is to 'range' the face at the ends of the stepped courses. This means using the spirit level as a straight edge, placed diagonally in line with the stepped brickwork as it racks back (see Figure 5.33). Gently tap the bricks to align them with the edge of the spirit level. Ranging the quoin will help to produce an accurate face plane in the finished wall.

Remember that quoins can be built to form external or internal corners in a cavity wall (see Figure 5.34). Whichever type of quoin is built, the same skills and checks apply to the stages of construction.

If the end of a wall is not built with a return corner, it may be designed as a 'stopped end'. This may require the cavity to be closed, which can be done by returning the inner leaf (usually built in block) into the outer leaf.

This type of design feature is used with some variations when forming openings in cavity walls. This will be covered in detail at Level 2.

▲ Figure 5.33 Using a spirit level diagonally to range the quoin

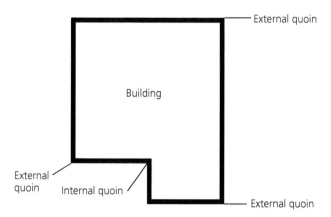

▲ Figure 5.34 External and internal quoin positions in a building

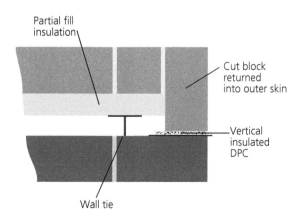

▲ Figure 5.35 Plan view of blockwork returned into the outer leaf

4.7 Building between the quoins

When all the checks show that the quoins have been constructed as accurately as possible, a string line can be attached to them as a guide to 'run in' the wall between them.

▲ Figure 5.36 String line attached to quoins

As mentioned in previous chapters, laying to the line is a skill that needs extensive practice to build speed and accuracy. Review the points about laying to the line in Chapter 4 and study the step-by-step illustrations.

Remember the key points when laying bricks or blocks to the line:

- Make sure the top arris of the brick or block is level with the line along its full length.
- Never lay the brick or block touching the line.

It is especially important when constructing face brickwork to ensure the perp joints are plumb. 'Eye' down the face of the wall to align the perps in alternate courses.

INDUSTRY TIP

When constructing high-quality face brickwork, some bricklayers use a T-square to line every fourth or fifth brick in a course with the perps below, marking the position with a pencil.

▲ Figure 5.37 Bricklayers laying to the line

4.8 Points of good practice

Until the mortar fully hardens, a wall is unstable and vulnerable to forces pushing against it, such as high winds. Even after the mortar has set, a single leaf of unsupported masonry can be pushed over if it is built too high.

Never build one leaf of masonry higher than six courses of block or 18 courses of brick in one operation. Build the two leaves of masonry in a cavity wall so that they support each other as the work progresses.

Develop good habits when forming mortar joints. It is especially important to form joints properly so that no gaps can form which could allow water to enter the wall. In freezing conditions, water standing in gaps in the wall will expand with considerable force and may cause the wall to split apart. This applies equally when forming bed joints and perp joints.

As mentioned in Chapter 4, some bricklayers have the habit of 'tipping and tailing' their perp joints to speed up the job. This means that instead of completely filling the joint, a small amount of mortar is placed on the front and back edges of the header face, leaving a gap in the finished joint. This is bad practice and weakens the finished wall.

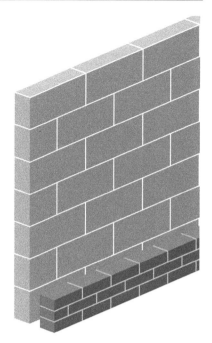

▲ Figure 5.38 A single leaf of blockwork seven courses high, unsupported by the brickwork

▲ Figure 5.39 Bricklayer incorrectly 'tipping and tailing' the perp joint

▲ Figure 5.40 Bricklayer correctly making a full perp joint

When forming the bed joint, some bricklayers use their trowel to create a deep furrow or groove in the middle of the mortar bed, with the intention of making it easier to adjust the position of the brick or block when it is laid. This can leave gaps under the brick or block when the mortar has hardened. It can also lead to mortar falling from the back edge of the outer leaf into the cavity.

Keeping the cavity clear of mortar droppings is extremely important. If such droppings build up at the bottom of the cavity and extend above the DPC, or build up on wall ties, they can form a bridge which allows moisture to enter the living or working space of the building.

4.9 Installing wall ties

Wall ties provide strength and stability. They are built into the bed joints of both leaves of a cavity wall and are carefully spaced. The position of wall ties is governed by building regulations. Study Figure 5.41 which shows the maximum horizontal and vertical measurements between wall ties.

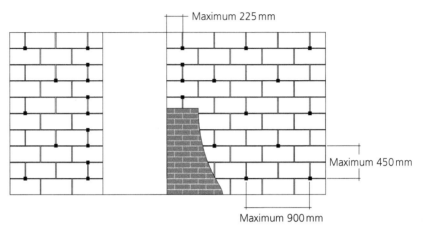

▲ Figure 5.41 Spacing of wall ties

INDUSTRY TIP

If the building design specifies a cavity wider than 75 mm, the maximum horizontal spacing of wall ties will be reduced from 900 mm to 750 mm to increase the stability of the structure.

Sometimes the bricklayer will decide to add more wall ties to provide enough strength in the cavity wall. It may be the case that positioning wall ties in accordance with the regulations from one end of the wall results in a gap in coverage at the other end of the wall. It is better to have too many ties than too few.

The position of wall ties across the cavity is important. As mentioned previously, the tie has a feature called a 'drip'. This must be positioned centrally across the cavity to make sure that moisture does not track across it and enter the living or working space of the building.

▲ Figure 5.42 Wall tie with drip centrally positioned in a cavity wall

Study the step-by-step guide to see how to install wall ties correctly.

Step by step
Installing wall ties correctly

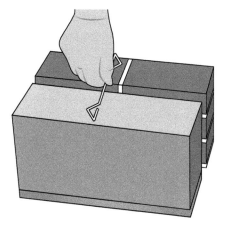

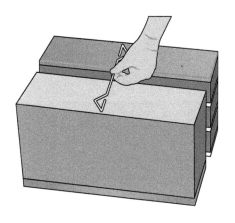

Step 1 Do not place the tie on the masonry before laying the mortar bed.

Step 2 Press the tie into the newly laid mortar bed.

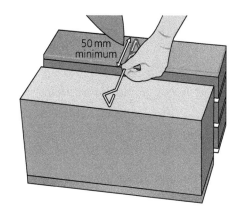

50 mm minimum

Step 3 Make sure the end of the tie is embedded at least 50 mm into the bed joint.

Step 4 Never try to push the tie into the mortar bed from the side at a later stage.

ACTIVITY

Wall ties must be kept clean and mortar droppings must be prevented from building up on them. Think of some ways of preventing mortar falling down the cavity.

Wall ties may seem like a relatively insignificant component of a cavity wall. However, they are very important in providing stability and strength. In the past, ties were used that were manufactured from materials that rusted and corroded. Many cavity walls deteriorated drastically, requiring expensive repair work. By installing wall ties correctly, the bricklayer plays an important part in constructing buildings that will last for many years.

4.10 Forming joint finishes

When the outer leaf of a cavity wall is constructed in facing brick, a joint finish must be formed for the bed and perp joints. The process of correctly forming joint finishes (or 'jointing') requires concentration and patience.

The standard of the jointing will affect both the appearance of the finished building and how weatherproof the masonry is. The tools used to form the joint finish are sometimes referred to as jointing irons. Ironing the joint will smooth and compress the surface of the mortar to seal it against the entry of moisture.

▲ Figure 5.43 Ironing the mortar joints to seal them

The most common joint finish is called half-round (or concave), which gives an attractive appearance when produced correctly. To create a good finish, joint all the perp joints first, followed by the bed joints. This reduces the number of untidy projections which can form where the perp and bed joints meet.

▲ Figure 5.44 'Mouse ears' or 'curtains' on brickwork jointing; these are small projections which occur when the horizontal and vertical joints cross each other

Look back at Table 4.4 in Chapter 4, which describes different types of joint finish.

Remember, it is the bricklayer's responsibility to ensure that the completed work meets the specification and has a good appearance without defects.

Test your knowledge

1 Which type of wall construction prevents moisture penetration when constructed correctly?

 a Straight

 b Solid

 c Return

 d Cavity

2 How many blocks are there in 1 m² of a half-brick-thick wall?

 a 10

 b 12

 c 15

 d 60

3 How many blocks are needed to build a half-brick-thick wall that is 3 m high by 3 m long?

 a 35

 b 50

 c 60

 d 90

4 Which item of PPE should a bricklayer wear when cutting lightweight blocks with hand tools?

 a Ear defenders

 b Dust mask

 c Rubber gauntlets

 d Knee pads

5 What feature is included in the design of a wall tie to prevent moisture travelling across it?

 a Drop

 b Strip

 c Drip

 d Stop

6 Which section of a building is described as the superstructure?

 a Below ground

 b Above ground

 c Below the ground floor

 d Above the ground floor

7 What is the minimum height above finished ground level for installation of a horizontal DPC?

   a 100 mm

   b 150 mm

   c 200 mm

   d 250 mm

8 If polythene DPC is used in a half-brick-thick wall, what is the minimum dimension that any joins should overlap by?

   a 100 mm

   b 150 mm

   c 200 mm

   d 220 mm

9 What is the gauged height measurement of a standard brick and a bed joint?

   a 65 mm

   b 75 mm

   c 90 mm

   d 100 mm

10 What is the main reason for preventing mortar from dropping into the cavity?

   a To save on materials

   b To make the cavity wall lighter

   c To prevent moisture bridging the cavity

   d To improve the appearance of the wall

CHAPTER 6

HEALTH AND SAFETY IN THE CONSTRUCTION INDUSTRY

INTRODUCTION

Working on building sites and in construction workshops can be rewarding. However, the work environment can present many potential hazards, which have unfortunately had a serious effect on the careers of many workers in the past. By continuously analysing work activities and formulating improved regulations, risks have been reduced and the workplace made safer.

Health and safety laws and regulations govern the work of a bricklayer on site. In this chapter we will look at how these laws and regulations benefit each worker and how you can apply them to contribute to workplace safety. Protecting yourself and those working with you will mean you can potentially look forward to a long and rewarding career in the construction industry.

LEARNING OUTCOMES

After reading this chapter, you should:
1   know about safety regulations, roles and responsibilities
2   know about identifying, reporting and dealing with hazards, accidents and emergencies
3   know how to handle materials and equipment safely
4   know about welfare in the workplace and the use of personal protective equipment
5   know about the safe use of access equipment and working at height.

1 SAFETY REGULATIONS, ROLES AND RESPONSIBILITIES

The introduction of effective health and safety legislation, combined with the efforts of construction workers, has made the workplace much safer in recent years. Everyone on site has a responsibility to make sure that high safety standards are maintained. That includes employers and workers like you.

Many workers have thought that an accident could never happen to them. The reality is that many still experience the consequences of an accident each year.

Accidents can have a devastating effect on individuals and their families. There can be a significant financial cost due to lost earnings and injury **compensation** claims. Workers can be prosecuted and lose their job if they are found to have broken safety laws.

KEY TERM

Compensation: something (usually money) awarded to someone in recognition of loss, suffering or injury

▲ Figure 6.1 Everyone on site must work to high safety standards

1.1 Health and safety regulations

The Health and Safety Executive (HSE) provides a lot of advice on safety and publishes numerous booklets and information sheets.

The HSE is a body set up by the government. It makes sure that the law is carried out correctly and has broad powers to ensure that it can do its job. It can carry out inspections, investigations and spot checks in the workplace. It can involve the police, examine anything on the premises and take things away to be examined.

An employer, employee, self-employed person (subcontractor) or anyone else involved with the building process can be taken to court for breaking health and safety **legislation**.

1.2 Health and safety legislation

The range of health and safety laws and regulations is wide. The following is a list of the main legislation that will affect a bricklayer working on site:

- Health and Safety at Work etc. Act (HASAWA) 1974
- Reporting of Injuries, Diseases and Dangerous Occurrences Regulations (RIDDOR) 2013
- Control of Substances Hazardous to Health (COSHH) Regulations 2002 (amended 2004)
- Construction, Design and Management (CDM) Regulations 2015
- Provision and Use of Work Equipment Regulations (PUWER) 1998
- Manual Handling Operations Regulations 1992 (amended 2002)
- Personal Protective Equipment at Work Regulations 2002
- Work at Height Regulations 2005 (amended 2007)
- Control of Noise at Work Regulations 2005
- Control of Vibration at Work Regulations 2005

KEY TERM

Legislation: a law or set of laws made by a government

▲ Figure 6.2 An HSE inspector

- Electricity at Work Regulations 1989
- Lifting Operations and Lifting Equipment Regulations (LOLER) 1998.

As a bricklayer, you are not expected to know every detail of this long list of legislation that applies to construction activities. However, you should be aware of the existence of each piece of legislation and its application in the workplace. In this chapter, we will focus on the key legislation that applies to a bricklayer's work.

1.3 Health and Safety at Work etc. Act (HASAWA) 1974

This legislation applies to all types of workplace and all personnel working there, including those delivering goods to the site. It also provides protection for those who might be affected by construction activities, such as members of the public who are nearby.

Employer's responsibilities under HASAWA

The employer's responsibilities are contained in sections 2 to 9 of the Act. An employer must take care of the health, safety and welfare at work of all their employees and provide:

- a safe working environment with adequate welfare facilities, including safe, secure access and exit points
- safe systems of work, including risk assessments and method statements (there is more information about these later in this chapter)
- clear systems for reporting hazards, accidents and near misses
- adequate instruction, training or supervision, including **site inductions** and **tool-box talks**
- **personal protective equipment (PPE)** free of charge, ensuring the appropriate PPE is used whenever needed
- facilities for the safe use, handling, storage and transport of components, materials and substances
- any required health and safety information, including health and safety law posters and public liability insurance details.

▲ Figure 6.3 Get to know how health and safety legislation applies to your work

KEY TERMS

Site induction: an occasion when someone is introduced to and informed about a new job or organisation

Tool-box talks: short meetings arranged at regular intervals at your work location on site to discuss safety issues; they give safety reminders and inform personnel about new hazards that may have recently arisen, for example, due to extreme weather

Personal protective equipment (PPE): all equipment (including clothing affording protection against the weather) which is intended to be worn or held by a person at work and which protects against one or more risks to a person's health or safety

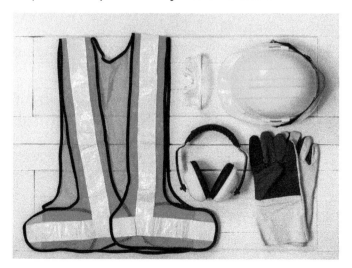

▲ Figure 6.4 PPE must be provided free of charge by the employer

A site induction is when someone is introduced to and informed about a new job or organisation. It will alert you to important details such as:

- specific hazards on site that you need to be aware of
- information about emergency assembly points, first-aid facilities and general welfare provisions
- site rules, including risks associated with the use of drugs and alcohol
- guidelines for keeping the site tidy and safe waste disposal.

INDUSTRY TIP

You will attend a site induction meeting when you arrive on site for the first time. Take a notebook with you and write down the key safety points that are discussed. Make a note of who the first aiders are.

▲ Figure 6.5 Induction meetings provide important information

Employee's responsibilities under HASAWA

A bricklayer's responsibilities on site are as follows:

- Always take care and work in a safe manner.
- Make sure that you do not put yourself or others at risk by your actions or inactions.
- Work in partnership with your employer regarding health and safety.
- Make full use of any equipment and safeguards (for example, PPE) provided by your employer.
- Do not interfere or tamper with safety equipment.
- Report hazards, accidents or near misses in accordance with laws and regulations.

1.4 Organisations providing information

While it may be difficult to remember all health and safety legislation details, it is important to know how to access this information when necessary. The relevant information is available online at the websites of these key organisations:

- Health and Safety Executive (HSE)
- Institute of Occupational Safety and Health (IOSH)
- British Safety Council (BSC)
- The Royal Society for the Prevention of Accidents (RoSPA)
- British Standards Institute (BSI)
- Royal Society for Public Health (RSPH).

Remember that health and safety information and instructions relating to specific materials, substances, tools and equipment can be obtained from the manufacturers of those items.

2 IDENTIFYING AND DEALING WITH HAZARDS AND REPORTING ACCIDENTS AND EMERGENCIES

Using the right tools for any construction task makes it easier to achieve success. Managing hazards successfully will also be made easier by using the right tools.

2.1 Risk assessments and method statements

The main tool used to identify and manage potential hazards is a document known as a risk assessment. A risk assessment can be split into three key stages:

1 Identify hazards and potential hazards in the workplace.
2 Assess the risk of harm likely to be caused by these hazards.
3 Establish measures to remove or minimise and control the risk.

Risk assessment template

Company name: M. Bloggs Ltd

Assessment carried out by: MB

Date of next review: 15/08/2021

Date assessment was carried out: 15/05/2021

What are the hazards?	Who might be harmed and how?	What are you already doing to control the risks ?	What further action do you need to take to control the risks ?	Who needs to carry out the action?	When is the action needed by?	Done
Brick stacks in work area	Bricklayer/General operative; Tripping hazard	Maintain good housekeeping	Remove waste regularly	General operative	Daily	✓

More information on managing risk: www.hse.gov.uk/simple-health-safety/risk/

Published by the Health and Safety Executive 09/20

▲ Figure 6.6 A risk assessment template

Reproduced by kind permission of HSE. HSE would like to make it clear it has not reviewed this product and does not endorse the business activity of Hodder Education.

INDUSTRY TIP

You can find more information on managing risk on the HSE website: www.hse.gov.uk/simple-health-safety/risk

A method statement can be used alongside a risk assessment. This gives a clear, uncomplicated sequence of work to achieve a specified task and can be used to record the specific hazards and potential hazards associated with the task. An employer will have a method statement written for all trade tasks that will be performed during a project.

Make sure you consult risk assessments and method statements before you start work on your assigned task.

2.2 Identifying and dealing with hazards

A construction site has many hazards and potential hazards that must be removed or suitably managed through well-designed risk assessments and method statements. These health and safety documents are only effective if all personnel on site use them appropriately. Be observant on site at all times and carefully consider the safety of those working near you.

Table 6.1 gives details of common hazards that must be managed on site.

▼ Table 6.1 Types of hazard

Type of hazard	How it can be caused
Slips, trips and falls	<ul><li>Trailing cables</li><li>Discarded brick or block pallet bands</li><li>Partially buried obstructions</li><li>Discarded waste materials</li><li>Unprotected excavations</li></ul> ▲ Loose brick bands can be a trip hazard
Manual handling injuries	<ul><li>Incorrect lifting techniques</li><li>Awkward-shaped loads</li><li>Obstructions in the path of travel when moving items</li></ul> ▲ Incorrect manual handling can cause injuries
Movements of plant, machinery and other vehicles	<ul><li>Personnel not wearing hi-vis clothing</li><li>Personnel out of line of sight of operator</li></ul> ▲ Be seen, be safe

▼ Table 6.1 Types of hazard (continued)

Type of hazard	How it can be caused
Electrocution	● Electrical items immersed in water ● Damaged cables ● Using poorly maintained or damaged equipment ▲ Cables can be damaged easily
Injuries from fires	● Unauthorised burning of waste ● Misuse of flammable liquids ● Poor storage of flammable items ▲ Store flammable liquids correctly

Hazardous substances

A bricklayer may come into contact with a number of hazardous substances and materials. Some give off harmful vapours or fumes which can be inhaled (breathed in). Other materials are described as irritants or corrosive and can cause damage to the skin or eyes. Some materials and substances that can be absorbed through the skin are toxic (poisonous).

The COSHH Regulations 2002 provide guidance on the safe use of potentially dangerous substances such as acids and chemical products.

Under HASAWA, it is the employer's responsibility to ensure the safe use, handling, storage and transport of components, materials and substances. However, the bricklayer must support the employer regarding the safe storage of **combustible** materials and hazardous chemical products on site.

Liquids which can catch fire, such as oil-based paints, thinners and oil, must be stored in a locked metal cupboard or shed. After using substances like these, always replace the cap or lid securely and return the product to secure storage. Hazardous chemicals should also be stored securely in a ventilated storage facility. Always follow the manufacturer's safety instructions regarding hazardous substances and use the correct PPE.

KEY TERM

Combustible: able to catch fire and burn easily

▲ Figure 6.7 Safe storage of hazardous substances on site

A material that was used a lot in the past is asbestos, the handling of which is covered by specific regulations. This has been recognised as a hazardous material that can cause a potentially fatal lung disease.

IMPROVE YOUR ENGLISH

Research asbestos and write a short report (less than one side of A4) stating what asbestos is made from and the name of the disease caused by it.

INDUSTRY TIP

Although discussing health and safety on site may make it sound like construction sites are extremely dangerous places, there is no need to be anxious about working in construction. Just follow the instructions and guidelines carefully and you will protect yourself and others from harm.

HEALTH AND SAFETY

If you come across what you think might be asbestos when working, do not disturb the material. Stop work immediately and inform your supervisor. Asbestos can be in the form of sheets of material or as fibres in material used to insulate pipes carrying hot gases or liquids.

Keeping things clean and tidy

Keeping your work area, and the site in general, clean and tidy ('good housekeeping') is important in managing risks.

Make it a habit to clear waste and debris from your work area regularly by moving it to designated waste areas on site, to avoid creating obstructions and trip hazards. Routes to fire escapes and access to fire extinguishers should never be blocked by waste materials.

When disposing of waste materials at height (for example, from a scaffold platform), never throw the materials from the working platform to the ground. Workers below could be struck and potentially suffer severe injury or even be killed. Purpose-made chutes are available to safely channel waste into a suitably placed skip at ground level. (There is more information about this later in this chapter.)

Cleanliness and personal hygiene are important in maintaining health and safety in the workplace. For example, if you have been handling materials such as lead (a soft, easily shaped metal which is toxic), washing your hands before eating will protect you from ingesting (swallowing) toxic materials. Hand washing will also protect against diseases such as leptospirosis (also known as Weil's disease). This is a serious disease spread by contact with urine from rats, which are often present on construction sites.

▲ Figure 6.8 Avoid accidents – keep your work area tidy

Signs and safety notices

Signs and safety notices in the workplace can help workers to manage potential hazards and reduce accidents and emergencies.

You should be familiar with a range of standardised safety signs used in construction. Look at Table 6.2 to see some examples.

▼ Table 6.2 Safety signs and notices

Type of sign	Description
Prohibition 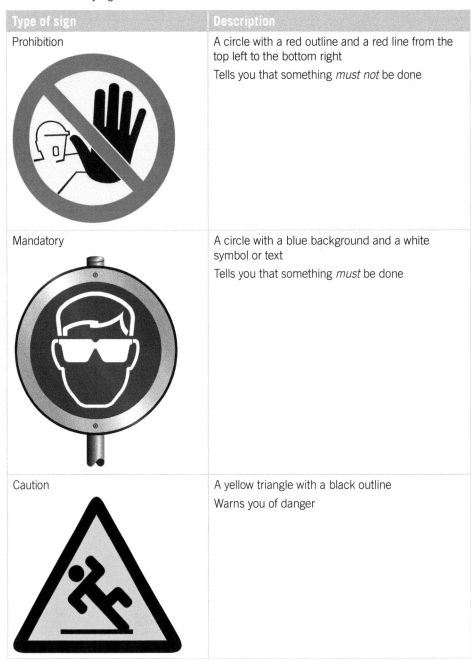	A circle with a red outline and a red line from the top left to the bottom right Tells you that something *must not* be done
Mandatory	A circle with a blue background and a white symbol or text Tells you that something *must* be done
Caution	A yellow triangle with a black outline Warns you of danger

▼ Table 6.2 Safety signs and notices (continued)

Type of sign	Description
Safe condition	A square or rectangle with a green background
	Shows directions to areas of safety and medical assistance in case of emergency
Supplementary	Square or rectangle with a white background
	Gives additional important information; usually used alongside 'safe condition' signs

ACTIVITY

Look for prohibition and mandatory signs in your workplace or training centre. List two different applications for each type of sign. (Hint: a common prohibition sign is 'No smoking'.)

2.3 Reporting accidents and emergencies

By making reports and keeping records of accidents on site, it is possible to see patterns that may be emerging, possibly due to bad habits and incorrect work practices. This provides the opportunity to make changes that will improve safety for all workers on site.

Keeping records of accidents and reporting them to the HSE is governed by RIDDOR 2013. These regulations state that employers must report to the HSE all accidents that result in an employee needing more than seven days off work.

Diseases identified on site and dangerous occurrences such as fires, gas leaks and security incidents must also be reported. A dangerous occurrence which has not caused an injury (a near miss) should also be reported because if it were to happen again, the consequences could be more serious. Steps should be taken to minimise the likelihood of a similar occurrence happening again. RIDDOR lists what are termed 'specified injuries' that must be reported.

ACTIVITY

Go to www.hse.gov.uk and enter 'specified injuries' into the search bar. Choose one of the injuries and suggest possible causes of that injury on site.

INDUSTRY TIP

The HSE website also gives specific details of reportable diseases and dangerous occurrences.

HSE Health and Safety Executive

Health and Safety at Work etc Act 1974
The Reporting of Injuries, Diseases and Dangerous Occurrences Regulations 1995

F2508 - Report of an injury

About you and your organisation

*Title   *Forename   *Family Name

*Job Title   *Your Phone No

*Organisation Name

Address Line 1   (eg building name)

Address Line 2   (eg street)

Address Line 3   (eg district)

*Town

County

*Post Code   Fax Number

*E-Mail

☐ Remember me   [?]

*Did the incident happen at the above address?   ☐ Yes   ☐ No

*Which authority is responsible fo r monitoring and inspecting health and safety where the incident happened?   ☐ HSE   ☐ Local Authority   [?]
Please refer to the help for guidance on the responsible authority

[Next]   [Form Preview]

Page 1 of 5

▲ Figure 6.9 An F2508 injury report form

Reproduced by kind permission of HSE. HSE would like to make it clear it has not reviewed this product and does not endorse the business activity of Hodder Education.

Whether or not a report to the HSE is required, records on site must be kept in an accident book to comply with regulations. These records can assist in planning future work activities to reduce the occurrence of accidents and may be used when legal matters arise related to an accident or emergency.

INDUSTRY TIP

Even minor injuries such as a cut finger should be recorded in an accident book on site. Do not be tempted to think 'it will be alright'. If the cut becomes infected later on, the record of when the original injury occurred could be useful.

Individual organisations and companies may have their own health and safety recording documentation, which a bricklayer should be familiar with. For example, on-site records can be used to provide information about personnel trained to give first aid.

The aim of first aid is to stabilise a patient for later treatment if required, so proper training of first aiders is vital. The injured person may need to be taken to hospital, or an ambulance or other emergency services may need to be called. The first aider must be able to communicate effectively with supervisors and on-site safety officers.

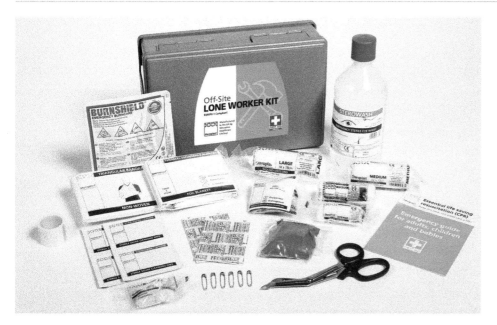

▲ Figure 6.10 A first-aid kit

Everyone on site must work hard to create a safe working environment. Accidents do not just affect the person who has the accident. Work colleagues or members of the public might also be affected, and so will the employer.

Some of the potential consequences that can have an effect on all site workers are:

- emotional trauma
- poor company image
- loss of production
- increased insurance costs
- closure of the site.

If you are near an accident or emergency situation on site, make the area safe if it does not put you in danger to do so, and get help. If necessary, call for a first aider. Make sure you or someone else calls the emergency services and report the matter immediately to your supervisor. Your supervisor will alert the site safety officer who will assess the situation carefully.

2.4 Fire safety

Fire safety relies on all personnel on site being aware of the potential hazards that could lead to an outbreak of fire. For a fire to exist, three things are needed:

- oxygen – a gas that occurs naturally in air
- heat – such as a spark or naked flame
- fuel – any material or substance that is combustible.

If all three of these things are present, a fire is unavoidable. If one of these things is missing, a fire cannot occur. This is often referred to as the 'fire triangle'.

▲ Figure 6.11 The fire triangle

Preventing the spread of fire

Maintaining a clean and tidy workplace is a key factor in helping to prevent fires from starting and spreading. Offcuts of wood left in piles, or paper and cardboard piled together, will provide fuel if a fire does start. Dispose of waste appropriately and store materials such as timber offcuts in storage racks or similar.

Take care never to have a naked flame near flammable liquids such as paint thinners or fuels such as petrol, and make sure that caps and lids on containers are closed securely.

Raising the alarm

It is vital that you know what to do if an emergency such as a fire occurs. If you discover a fire, raise the alarm immediately. You should know what the alarm sounds like, for example, a bell or a siren.

Make sure you know where the fire assembly point is in advance of a possible evacuation and make a note of the route from your work location. Remember: if you are not at the fire assembly point when an alarm is sounded, others will have to look for you, which could put them at risk.

▲ Figure 6.12 Fire assembly point

INDUSTRY TIP

Familiarise yourself with fire escape routes and alarm signals on site, especially if you are a new starter.

ACTIVITY

Walk from where you work to your designated fire assembly point. Time how long it takes at a normal walking pace.

If you are required to leave a building and proceed to the fire assembly point, leave belongings behind and do not return to the building until the appointed fire marshal tells you it is safe to do so.

Fire extinguishers

To contribute to fire safety, you need to know the location of fire extinguishers and fire blankets in your workplace and know which equipment can be used on different fires.

Fires are classified according to the type of fuel that is burning. Study Table 6.3 to see which type of extinguisher can be used for each type of fire.

▼ Table 6.3 Fire classifications and extinguishers

Classification of fire	Type of fuel	Type of extinguisher to use
A	Wood, paper, textiles	Water, foam, dry powder
B	Flammable liquids	Foam, dry powder, CO_2
C	Flammable gases	Dry powder, CO_2
D	Flammable metals	Specially formulated dry powder
Electrical	–	CO_2, dry powder
F	Cooking oils	Fire blanket

An additional type of fire extinguisher called a P50 is certified to deal with a wide range of fire classifications.

It is important to use the correct extinguisher to deal with the different types of fire, as using the wrong one could make the danger much worse. For example, using a water extinguisher on an electrical fire could lead to the user being electrocuted. Although all fire extinguishers are red, they each have a different coloured label to identify their contents, as shown in Table 6.4.

▼ Table 6.4 Types of fire extinguisher

Type	Label colour	Image
Water	Red	
Foam	Cream	

▼ Table 6.4 Types of fire extinguisher (continued)

Type	Label colour	Image
Carbon dioxide (CO_2)	Black	
Dry powder	Blue	

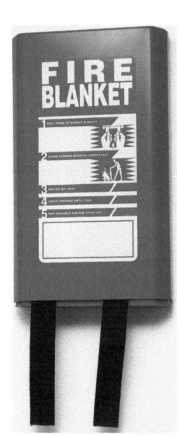

▲ Figure 6.13 Fire blanket

Fire blankets are safety devices designed to extinguish small fires before they spread. They consist of a sheet of fire-retardant material that is placed over a fire to smother it.

3 HANDLING MATERIALS AND EQUIPMENT SAFELY

Health and safety laws and regulations are designed to cover every activity on a construction site. There is specific legislation designed to protect workers when handling materials and using certain types of equipment.

3.1 Manual Handling Operations Regulations 1992 (amended 2002)

Many workplace injuries are a result of manual handling.

Within reason, employers and training providers must arrange for personnel to avoid manual handling if there is a possibility of injury. If manual handling cannot be avoided, then they must minimise the risk of injury by conducting an appropriate risk assessment. Remember, pushing or pulling an object still comes under the Manual Handling Operations Regulations.

Using the correct lifting techniques and the proper equipment is essential to avoid injury. Consider the following points when moving materials or components manually:

- Think about the weight of the item to be moved. Is it too heavy or awkward for one person to handle? Will assistance be needed?
- Is there a lifting aid or a mechanical means of moving the load?
- Make sure the path along which the load is to be moved is clear of obstructions.
- Avoid twisting your body or reaching too much when moving the load.

For safety and efficiency, most sites will have some means of moving heavy items mechanically, such as a forklift or crane. If items cannot be moved by mechanical means and must be moved manually, methods have been developed to move and handle components and materials efficiently and in a way that reduces the risk of injury. One recommended method is called **kinetic** lifting.

> **KEY TERM**
>
> **Kinetic:** relating to, caused by or producing movement

▲ Figure 6.14 Safe kinetic lifting technique

INDUSTRY TIP

Build a good relationship with the operators of forklifts and cranes. You will not only improve the safety of transport operations, but you will also increase your productivity if you work well together.

To employ the kinetic lifting method, keep the following points in mind:

- Always lift with your back straight.
- Keep your elbows in, knees bent and feet slightly apart.
- When placing the item, be sure to use your leg muscles, bending your knees.
- Beware of trapping your fingers when stacking materials.
- Place the item on levelled, carefully spaced bearers if required.

For very heavy items, get assistance from one or more helpers. Assign one person in the team to be in charge, and make sure that lifting is done in a cooperative way.

INDUSTRY TIP

An item that is not too heavy for one person to lift may still be awkward to move on your own without causing injury. For example, a long length of timber should be moved with one person at each end, to avoid causing injury to others on site.

3.2 Provision and Use of Work Equipment Regulations (PUWER) 1998 (amended 2018)

These regulations apply to anyone who has responsibility for the safe use of work equipment, such as managers and supervisors. The HSE website states that it applies to:

'... people and companies who own, operate or have control over work equipment. PUWER also places responsibilities on businesses and organisations whose employees use work equipment, whether owned by them or not.'

ACTIVITY

Visit www.hse.gov.uk and enter 'work equipment' in the search bar. On the page about PUWER, look for the section titled 'What is work equipment' and write your own version of the details there. Discuss with another person in your learner group if any workers on site are exempt from applying PUWER.

▲ Figure 6.15 Regular safety inspections of equipment are essential

Under these regulations:`

- Regular safety inspections of equipment are essential.
- Work equipment must only be used by workers who have been properly trained to the required level of competence.
- Relevant information on the safe use of the equipment must be provided.
- Equipment must only be used for its intended purpose.

3.3 Lifting Operations and Lifting Equipment Regulations (LOLER) 1998 (amended 2018)

These regulations place duties on people and companies who own, operate or have control over lifting equipment.

All lifting operations must be:

- properly planned by a competent person
- appropriately supervised
- carried out in a safe manner.

All lifting equipment must be:

- fit for purpose
- appropriate for the task
- suitably marked for record keeping
- subject to a thorough periodic examination.

HEALTH AND SAFETY

Under PUWER, abrasive wheels used for cutting or grinding can only be changed by someone who has received training to do this. Wrongly fitted wheels can fly apart and cause serious injury.

▲ Figure 6.16 Using a scissor lift at height

The equipment used for lifting operations will also be subject to PUWER. As a bricklayer, you will likely use simple lifting aids when moving materials and components, such as wheelbarrows and brick tongs, or perhaps a mechanical lifting aid such as a pallet truck on level surfaces.

HEALTH AND SAFETY

Even when using a lifting aid such as a wheelbarrow or brick tongs, you are performing a form of manual handling. Remember, pushing and pulling a load in any way comes under the Manual Handling Operations Regulations discussed previously in this chapter. Different regulations often link with each other.

▲ Figure 6.17 Wheelbarrow

▲ Figure 6.18 Brick tongs

Remember that even when using a lifting aid, it is still important to assess the weight of the load, how awkward it is to move and handle, and whether assistance is needed for a safe lifting operation. Plan the lifting operation and make sure the path of movement has no obstructions.

3.4 Electricity at Work Regulations 1989

Work involving electrical equipment is regulated by the Electricity at Work Regulations 1989. Electricity is an everyday part of our lives, so it can be easy to take it for granted as an energy source.

▲ Figure 6.19 Many electrical substations can be found around towns and cities – they play an important role in transmitting electricity

Never forget that electricity can be dangerous if it is not treated with respect, and it can quickly cause severe injury or death. Make sure that you are properly trained before using electrical equipment in the workplace. Never dismantle or adjust electrical equipment of any kind. Leave it to someone who is authorised, trained and competent.

When working with or near equipment powered by electricity, be aware of the types of danger that can arise:

- burns and electric shock, which depending on the **voltage** can kill you
- faults in equipment or wiring which can cause a fire
- electrical sparks caused by faulty equipment which can cause flammable gas to explode.

Voltages

Different voltages are used, depending on the equipment and circumstances.

In our homes, the usual voltage supplied for domestic appliances is 230 V, commonly referred to as 240 V. The difference between these two voltage figures is because voltages are referred to as 'nominal', which means they can vary slightly.

230 V is often used in workshop environments to power hand tools and fixed machines. Protection for users is provided by a residual current device (RCD). This will disconnect the supply quickly if a fault or unsafe condition occurs. Heavier equipment or machinery may require a 410 V supply, which you will hear referred to as 'three phase'.

On site, a lower voltage of 110 V is recommended, since lower voltages are safer. A piece of equipment called a 'transformer' is used to reduce a 230 V supply to the safer voltage of 110 V.

Each voltage level is colour coded, to make safe identification easier:

- 110 V – yellow
- 230 V – blue
- 410 V – red.

Wiring

Cables must be made of a material that is a good **conductor** of electricity. Copper wire is used as a conductor in the cable connecting a power tool to an electricity source. Three separate **insulated** wires are contained in an outer coating, and these wires are colour coded as follows:

- The live wire is brown and conducts electricity to the powered device.
- The neutral wire is blue and completes the electrical circuit back to the power source.
- The earth wire is green and yellow striped and provides a path for the electrical current if the appliance is faulty or damaged.

KEY TERMS

Voltage: the amount of potential energy between two points in an electrical circuit, expressed as 'volts' or 'V'

Conductor: a material through which electricity can flow freely

Insulated: covered in a material through which electricity cannot flow freely

▲ Figure 6.20 Colour-coded copper wires in a cable

When a plug is fitted to a cable, the colour-coded wires must be connected to the correct pins.

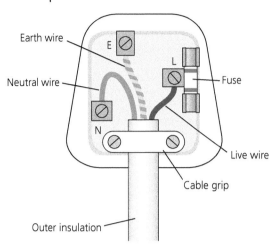

▲ Figure 6.21 A correctly wired plug

Battery-powered tools

Battery-powered tools are safer than mains-powered tools, since they operate at lower voltages. They are available in a wide variety of voltages, from 3.6 V for a small powered screwdriver all the way up to 36 V for a large masonry drill.

ACTIVITY

Visit a tool supplier's website and search for battery-powered tools (they may be referred to as 'cordless'). List four types of battery-powered tool available and note the voltages.

▲ Figure 6.22 Battery-powered drill

Safety precautions

Always check your power tools before use. If a fault is identified, inform a supervisor immediately. The tool will need to be repaired and should not be used. It should be removed from the work area, put in a secure location and clearly labelled 'Do not use'.

Study Table 6.5 to see the safety checks required when using electrical tools and equipment.

▼ Table 6.5 Safety checks on electrical tools and equipment

Item	What to check for	Action to take
Cables and plugs	Are there any signs of damage? Have they been repaired in the past? (Insulation tape may be hiding a damaged cable or plug.)	If there is any sign of damage, remove the tool from use until it is properly repaired.
Electricity supply leads	Are there any signs of damage? Have they been repaired in the past? (Insulation tape may be hiding a damaged supply lead.)	If there is any sign of damage, remove the tool from use until it is properly repaired.

▼ Table 6.5 Safety checks on electrical tools and equipment (continued)

Item	What to check for	Action to take
Electricity supply leads	Are the supply leads creating a trip hazard?	Make sure that supply leads on the ground are protected from traffic or personnel crossing over them. If possible, use cable hangers to run supply leads overhead, out of harm's way.
Power tools and equipment	Are there any signs of damage to the casing of the power tool?	Plastic casings on power tools provide high levels of protection against electric shock when they are undamaged (often referred to as 'double insulated'). If a casing is cracked, the tool should not be used.
PAT sticker	Is the PAT sticker up to date?	PAT (portable appliance testing) is carried out by a trained and qualified electrician to check the safe electrical condition of tools and equipment. A sticker is placed on the tool after it has been tested. Tools that do not pass PAT must be taken out of use.

▲ Figure 6.23 Cable protection

Always fully unroll an extension lead before using it. If you leave the cable coiled up during use, especially if it has a casing, it can overheat and cause a fire.

After use, store cleaned power tools and equipment in their carrying boxes in a clean and secure location, making sure that all the parts and accessories are present and undamaged. Cables should be wound onto reels or neatly coiled to avoid them becoming tangled or damaged.

▲ Figure 6.24 Cable reel

3.5 Control of Vibration at Work Regulations 2005

'Vibration white finger', also known as hand-arm vibration syndrome (HAVS), is caused by using tools that create a lot of vibration, such as handheld breakers or hammer drills.

To reduce the risk of this occurring:

- Limit exposure to vibration by using these tools for the shortest time possible.
- Alternate the work task involving vibration with other work.
- Always use the right tool for the job.
- Make sure cutting edges are sharp to make the tool as efficient as possible.
- Use appropriate PPE including vibration reducing gloves.
- Keep warm and dry to encourage good blood circulation, and massage fingers during breaks from work.
- Avoid gripping the tool too tightly or forcing the tool into the work piece trying to speed up the job. Let the tool do the work.

▲ Figure 6.25 Do not use power tools for long periods without a break

ACTIVITY

Research HAVS on the HSE website and list the symptoms and injuries that it can cause.

4 WELFARE IN THE WORKPLACE AND THE USE OF PERSONAL PROTECTIVE EQUIPMENT

The welfare of workers on site is protected individually and collectively by regulations that focus on the effective planning and management of construction projects, with a focus on health and safety considerations.

These regulations are designed to reduce the risk of harm both to those who work on the construction and those who use the building throughout its existence.

During the construction phase, protection for individual workers is provided by specific regulations governing the use of personal protective equipment (PPE).

4.1 Construction, Design and Management (CDM) Regulations 2015

The welfare of workers on site is cared for under the CDM regulations, by the provision of a range of specified facilities. Table 6.6 gives details of what is required under the CDM regulations.

▼ Table 6.6 Welfare facilities that must be provided by employers

Facility	Regulation requirement
Clean drinking water 	Clean drinking water must be provided or made available. There should be appropriate signs showing where drinking water is available. Unless the supply of drinking water is from a water fountain, cups should be provided.
Toilets 	Enough suitable toilets should be provided or made available. Toilets should be: ● adequately ventilated and lit ● maintained in a clean condition ● separate for male and female.
Washing facilities 	Enough washing facilities must be available, including showers if required by the nature of the work. These facilities should be: ● in the same place as the toilets and near any changing rooms ● supplied with clean hot (or warm) and cold running water, soap and towels ● separate for male and female, unless the area is for washing the hands and face only.
Changing rooms and lockers 	If operatives must wear special clothing and if they cannot be expected to change elsewhere, changing rooms must be provided or made available. Separate rooms for males and females should be available where necessary. There should be facilities for operatives to dry clothing, with seating where necessary. Lockers should be provided.
Rest rooms or rest areas 	There should be enough tables and seating with backrests for the number of operatives using them at any one time. There should be facilities for preparing, heating and eating food. There must be provision to boil water.

Workers on site have a responsibility to look after the safety and welfare of themselves and others by acting appropriately regarding the use of alcohol and drugs during working hours.

Remember, alcohol and drugs can have a serious effect on your performance and behaviour in the workplace, reducing your perception of risk, causing a loss of concentration and impairing your balance. Remaining on a construction site under the influence of alcohol or drugs is dangerous, and a worker in that condition will very likely be excluded from the site and could face serious consequences. (Keep in mind also that some medications can affect a worker in an undesirable way.)

Put the health, safety and welfare of yourself and others working near you at the forefront of everything you plan for and work towards on site.

4.2 Control of Noise at Work Regulations 2005

The aim of these regulations is to ensure that workers' hearing is protected from excessive noise at their place of work. Exposure to noise caused by power tools or machinery can cause permanent hearing loss. Damage to hearing can mean a worker loses the ability to understand speech, keep up with conversations or use the telephone.

A single loud noise or persistent lower levels of noise can both cause damage to your hearing that cannot be repaired. An indication of when you should wear ear defenders or use ear plugs is when you have to raise your voice to carry out a normal conversation two metres apart from someone else.

HEALTH AND SAFETY

Research 'tinnitus' online and find out the symptoms and what causes it. Is there a cure?

ACTIVITY

Go to www.hse.gov.uk/noise/video/hearingvideo.htm and watch 'The Hearing Video' to see how important it is to protect your hearing.

▲ Figure 6.26 Ear defenders

▲ Figure 6.27 Ear plugs

INDUSTRY TIP

Since ear plugs are such small items of PPE, it is easy to carry a set with you at all times. If a worker near you starts using a noisy tool, you are already prepared to protect yourself.

Employers are required to take action to reduce noise exposure and provide appropriate PPE when noise exceeds specified levels.

Actions to minimise the risk of hearing damage to workers include:

- using quieter equipment or a quieter process
- using isolation screens, barriers and absorbent materials to prevent noise travelling
- improving working techniques to reduce noise levels
- limiting the time workers spend in noisy areas.

Employers must make sure that workers have appropriate information, instruction and training, and provide appropriate supervision. As a bricklayer, you have the responsibility to make use of any noise-control measures provided to you.

4.3 Personal Protective Equipment (PPE) at Work Regulations 2002

Employers must provide appropriate PPE without charge to employees, including agency workers where they are legally recognised as employed by the contractor. You must wear the PPE that your employer provides, look after it while using it, report any damage to it and store it correctly.

If PPE is not cared for properly and stored correctly, it can be damaged and may not provide adequate protection to the user. For example, ultraviolet light from the sun will cause the plastic from which a hard hat is made to deteriorate over time. Some pens and paints that might be used to write on a hard hat can also damage the plastic. Check the condition and date of manufacture and replace any item that is damaged or out of date.

Table 6.7 shows the types of PPE used in the workplace.

▼ Table 6.7 Types of PPE used in the workplace

PPE	Description/Use
Hard hats/safety helmets 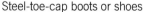	Hard hats must be worn when there is danger of: ● striking your head on overhead obstructions ● being hit by objects falling from above. Most sites insist on hard hats being worn at all times. They must be adjusted to fit your head correctly and must not be worn back to front!
Steel-toe-cap boots or shoes 	Steel-toe-cap boots or shoes are worn at all times on site to protect the feet from crushing by heavy objects. Some safety footwear has additional insole protection, to help prevent nails going up through the foot.

▼ Table 6.7 Types of PPE used in the workplace (continued)

PPE	Description/Use
Ear defenders and plugs	Your hearing can be permanently damaged by a single loud noise or persistent lower noise levels. Hearing protection in the form of ear defenders or ear plugs is an important item of PPE.
High visibility (hi-vis) jackets	Hi-vis clothing is essential on site, so that other people can see you easily. Plant or machinery being used nearby or moved around a site is a potential danger to all workers.

▼ Table 6.7 Types of PPE used in the workplace (continued)

PPE	Description/Use
Goggles and safety glasses 	Dust and flying debris are constant features of a construction site. Your eyes can be damaged easily. Wearing goggles or safety glasses is a simple way to protect them.
Dust masks and respirators 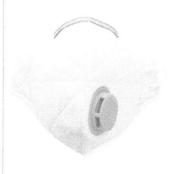 	Dust can damage your lungs and cause serious illnesses that can be life threatening. A dust mask gives vital protection when correctly fitted. Note: a respirator is used to filter out hazardous gases. Respirators are rated P1, P2 and P3 to show the level of protection they provide. Equipment that protects your breathing is referred to as RPE (respiratory protective equipment).
Knee pads 	Knee pads provide essential protection for the knee joints when kneeling for extended periods.

▼ Table 6.7 Types of PPE used in the workplace (continued)

PPE	Description/Use
Gloves 	Gloves protect against cuts, abrasions, skin irritations, chemical burns and splinters. Different types of gloves are available to suit different work tasks, providing a good grip and protecting the fingers.
Sunscreen	A real risk, especially in the summer months, is sunburn. Use sunscreen as PPE against burning. Excessive exposure to the sun can cause skin damage that may lead to cancer.

INDUSTRY TIP

Waterproof clothing and overalls can be used as protection in extreme working conditions.

ACTIVITY

Find a supplier of PPE online and note the cost of each item of PPE in Table 6.7 (there is no need to do extensive research – just use the cost of the first item you come across in each instance). Add up the total cost for a full set of items.

Think about the cost to your employer if it must provide PPE for each worker on site. Look after your PPE – it is a valuable provision!

▲ Figure 6.28 A site safety sign showing the PPE required to work in this area

5 THE SAFE USE OF ACCESS EQUIPMENT AND WORKING AT HEIGHT

Falls from height are one of the biggest causes of fatalities and major injuries on construction sites. Any place where, if there were no precautions in place, a person could fall a distance liable to cause personal injury can be termed 'working at height'.

Employing safe methods of work and using suitable access equipment correctly are vital factors in keeping yourself and others safe when working at height. All equipment should be checked for safe condition before use and inspection tags and notices should be up to date. Never take short cuts!

▲ Figure 6.29 Workers wearing safety harnesses on an aerial access platform

5.1 Work at Height Regulations 2005 (amended 2007)

These regulations give clear guidance to employers regarding the planning and preparation required for working at height. The following points must be carefully considered:

- Avoid work at height where it is reasonably possible to do so.
- Where work at height cannot be avoided, minimise the distance and consequences of a fall by using the right type of equipment.
- Do as much work as possible from the ground.
- Make sure workers can get safely to and from where they work at height.

If you must work at height, think about these points before you start work:

- Make sure the equipment is stable and strong enough for the job, maintained and inspected regularly, especially after bad weather. If there are broken, damaged or missing components, do not use it.
- Make sure you do not overload or overreach when working at height.
- Be responsible and provide protection from falling objects for people below you. Make sure there are barriers to prevent materials falling from the working platform and keep your work area tidy.

▲ Figure 6.30 A cherry picker can provide a safe working platform

5.2 Types of access equipment and safe methods of use

A risk assessment and method statement must be produced before work at height begins. These documents will assist in choosing the right type of access equipment for the job.

Ladders

Ladders are mainly used for access onto an elevated working platform, but the regulations allow for their use as a working platform for light work over short periods. The top of the ladder must be secured against something firm and stable, not something flexible such as plastic guttering.

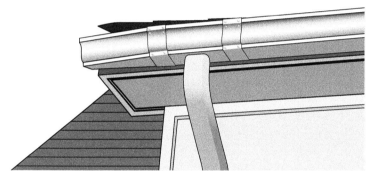

▲ Figure 6.32 Resting ladders on plastic guttering can cause it to bend and break

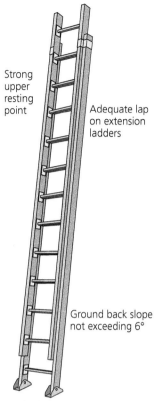

Strong upper resting point

Adequate lap on extension ladders

Ground back slope not exceeding 6°

Ground side slope not exceeding 16°, clean and free of slippery algae and moss

▲ Figure 6.31 Using a ladder correctly

IMPROVE YOUR ENGLISH

Go to www.hse.gov.uk and search for 'safe use of ladders and stepladders'.

In your own words, write a summary of the guidance on when you can use a ladder as a working platform.

A ladder should be set at an angle of 75° to the ground. Using a ratio of 1:4 ensures you are placing your ladder at the correct angle; the distance between the wall and the base of the ladder should be one quarter of the ladder's height.

INDUSTRY TIP

Take your time and be observant when setting up a ladder. Many workers have rushed to get on with a job and fallen because they failed to notice a problem until it was too late.

There are ladders designed for specific purposes, such as accessing a roof surface without causing damage to it. Always use a ladder that is designed for the job.

Some ladders are fixed in length (pole ladders) and others can be extended (extension ladders). When using an extension ladder, leave sufficient overlap between the two sections to maintain the strength and stability of the whole unit.

Working from the side can make stepladders unstable. Do not over-reach

Don't stand on the top three steps

Stepladder is fully open

Locked open firm and level on the ground

Stepladders

Stepladders are designed for short-term light work when used as a working platform.

Be careful not to reach out too far from the side of the stepladder, causing it to topple sideways.

If the stepladder has a shelf at the top, do not use this as a step. This is designed to carry light loads, such as tins of paint, or as a rest for placing small tools.

Always check ladders and stepladders carefully before using them. Check the stiles (the uprights) and rungs for damage such as splits or cracks.

INDUSTRY TIP

Do not use painted ladders or stepladders because the paint could be hiding damage.

▲ Figure 6.33 Using a stepladder correctly

▼ Table 6.8 Using ladders and stepladders safely

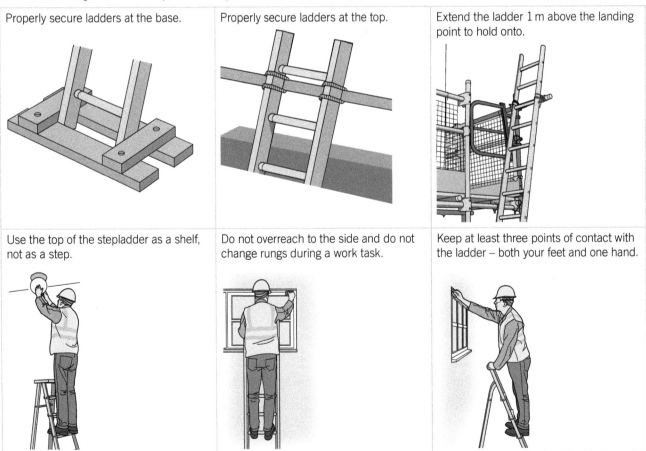

Properly secure ladders at the base.	Properly secure ladders at the top.	Extend the ladder 1 m above the landing point to hold onto.
Use the top of the stepladder as a shelf, not as a step.	Do not overreach to the side and do not change rungs during a work task.	Keep at least three points of contact with the ladder – both your feet and one hand.

Trestles

Trestles form an access system used for relatively short durations. This equipment is not designed to carry heavy loadings of materials, so a bricklayer using it will need to limit the weight placed on the platform when stacking materials for a work task. Trestle platforms must have edge protection to prevent tools and materials from falling off the working platform.

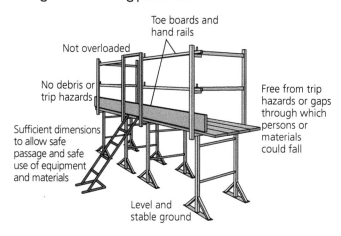

▲ Figure 6.34 Trestles used as a working platform

Hop-up

A hop-up is a low working platform that gives stable access for short-duration tasks. It will rarely be used by a bricklayer, except perhaps for tasks such as local small-scale repairs or alterations.

Mobile tower scaffold

Mobile tower scaffolds are manufactured from galvanised steel or lightweight aluminium alloy. They must be assembled by competent operatives in accordance with the manufacturer's instructions.

They should always have guard rails and toe boards fitted when in use, to prevent people and materials accidentally falling from the working platform.

Pay attention to these important points when using a mobile tower scaffold:

- Read and follow the manufacturer's instructions.
- Use the equipment for its designed purpose.
- Do not exceed the maximum height given in the manufacturer's instructions.
- Do not overload the working platform.
- Make sure the wheels of the scaffold are on a firm surface (some scaffolds have feet instead of wheels).
- Unload the working platform and, if necessary, reduce the height before moving the scaffold.
- Do not move a tower scaffold if workers are standing on it.
- If a trap door is fitted for access to the working platform, make sure it is closed after personnel pass through it.

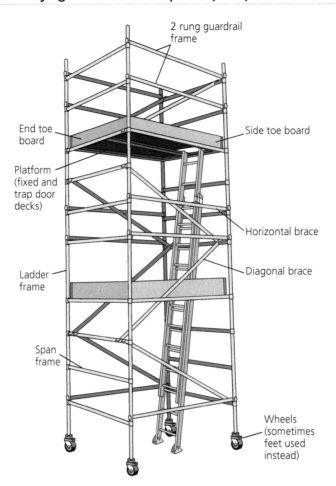

2 rung guardrail frame

End toe board

Side toe board

Platform (fixed and trap door decks)

Horizontal brace

Ladder frame

Diagonal brace

Span frame

Wheels (sometimes feet used instead)

▲ Figure 6.35 A tower scaffold and its parts

▲ Figure 6.36 Safe scaffold being set up

Tubular scaffold

Tubular scaffold is erected by specialist scaffolding companies who employ trained and highly competent operatives. It is assembled using steel tubes and clips which bolt tightly together in a specified pattern.

Some types of tubular scaffold are designed to be assembled using specially shaped cups on the upright tubes, which allow horizontal tubing to be slotted into them and wedged together tightly.

Never alter tubular scaffold yourself. There can be a temptation to remove handrails to allow materials to be placed more conveniently on the working platform by forklift or crane. However, this would be dangerous to yourself and others working on the scaffold.

When stacking materials on tubular scaffold, make sure the load is spread evenly to distribute the weight carried by the structure.

Tubular scaffold has been in use for many years, with refinements in design to improve safety. Two types of tubular scaffold have been used:

- Independent tubular scaffold transmits the loadings carried by it to the ground.
- Putlog tubular scaffold (rarely used now) transmits the loadings partly to the ground and partly to the building under construction.

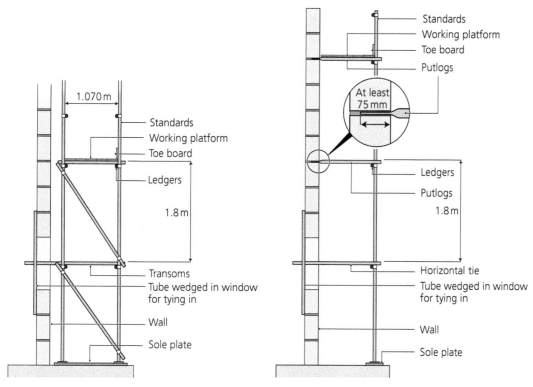

▲ Figure 6.37 Independent tubular scaffold

▲ Figure 6.38 Putlog tubular scaffold

ACTIVITY

Visit YouTube and search for Sprout Labs' 'Scaffolding Training Video' (www.youtube.com/watch?v=veF4uSUtrEY).

This video shows how trained scaffolders reduce the risks associated with working at height. Watch the video then create a poster highlighting the key points.

Scaffold must be erected and used so that it satisfies the following requirements:

- Handrails, brick guards, toe boards and working platforms must have no gaps along their length.
- Systems and equipment for lifting materials to working-platform level must comply with safety regulations.
- Systems and equipment for removing waste and debris from the elevated working platform must be safe. Never throw waste materials off the working platform of a scaffold. Use a debris chute or place the materials securely on a pallet or in a container for controlled removal by forklift or crane.

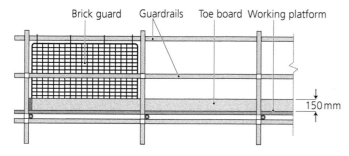

▲ Figure 6.39 Elements of a safe working platform

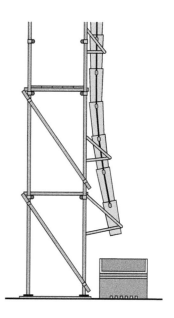

▲ Figure 6.40 A debris chute attached to scaffolding

Test your knowledge

1 What do the letters HSE stand for regarding safety inspections?

   a   Health and Standards Executive

   b   Health and Safety Environment

   c   Health and Standards Environment

   d   Health and Safety Executive

2 When a worker arrives on site for the first time, what type of meeting should they attend to learn about safety matters?

   a   Introduction

   b   Induction

   c   Information

   d   Interaction

3 Which type of safety sign is triangular and coloured yellow and black?

   a   Mandatory

   b   Prohibition

   c   Caution

   d   Supplementary

4 Which regulations specifically apply to the reporting of injuries?

   a   RIDDOR

   b   COSHH

   c   PUWER

   d   LOLER

5 Which voltage of electricity is used to make power tools safer on site?

   a   110 V

   b   230 V

   c   240 V

   d   410 V

6 What colour is the live wire in an electrical cable?

   a   Blue

   b   Green

   c   Brown

   d   Yellow

7 What do the letters PAT stand for regarding inspection of electrical equipment?

   a   Power appliance testing

   b   Portable appliance testing

   c   Power application testing

   d   Portable application testing

8 Which legislation is specifically designed to protect workers from HAVS?

  a Control of Noise at Work Regulations 2005

  b Manual Handling Operations Regulations 1992 (amended 2002)

  c Control of Vibration at Work Regulations 2005

  d Provision and Use of Work Equipment Regulations 1998

9 At what angle to the ground should a ladder be set?

  a 60°

  b 65°

  c 70°

  d 75°

10 Which part of a scaffold prevents objects from falling over the edge of the working platform?

  a Hand rail

  b Standard

  c Toe board

  d Putlog

Test your knowledge answers

CHAPTER 1

1 b – Elements
2 a – Contract document
3 d – North
4 b – 1:50 or 1:100
5 d – Heat transfer
6 a – Pad
7 b – 10°
8 d – Rafter
9 b – Linear
10 c – Professional

CHAPTER 2

1 c – Site plan
2 a – Below ground-floor level
3 b – Profiles
4 c – TBM
5 d – 90°
6 a – 1:5 or 1:10
7 c – Building line
8 b – Laser level
9 c – Screed
10 d – The longest side

CHAPTER 3

1 d – Schedule
2 b – 200 mm
3 c – Section
4 d – 440
5 c – Retarder
6 c – Stopped end
7 b – 100 mm
8 a – 6
9 b – EML
10 c – Half-round

CHAPTER 4

1 a – Elevation
2 d – 60
3 b – Scutch hammer
4 c – To spread loadings evenly
5 b – 102.5
6 b – ±3 mm
7 d – To establish the correct bond
8 d – Reverse
9 b – Broken
10 c – Pockets left in a wall to form a junction

CHAPTER 5

1 d – Cavity
2 a – 10
3 d – 90
4 b – Dust mask
5 c – Drip
6 d – Above the ground floor
7 b – 150 mm
8 a – 100 mm
9 b – 75 mm
10 c – To prevent moisture bridging the cavity

CHAPTER 6

1 d – Health and Safety Executive
2 b – Induction
3 c – Caution
4 a – RIDDOR
5 a – 110V
6 c – Brown
7 b – Portable appliance testing
8 c – Control of Vibration at Work Regulations 2005
9 d – 75°
10 c – Toe board

Improve your maths answers

PAGE 9

- Foundation excavation: 7.5 days
- Masonry to ground floor: 7.5 days
- Installation of floor slab: 5 days

PAGE 31

1 Volume of concrete needed: $4 \times 0.6 \times 0.3 = 0.72$ m³
2 Volume of concrete needed: $6.5 \times 0.4 \times 1.2 = 3.12$ m³

PAGE 32 TOP

1 Area of the wall: $5.5 \times 0.9 = 4.95$ m²
  Number of bricks needed: $4.95 \times 60 = 297$
2 Area of the wall: $9 \times 0.45 = 4.05$ m²
  Number of blocks needed: $4.05 \times 10 = 40.5$

PAGE 32 BOTTOM

1 Sides 1 and 2 added together: $2 \times 6 = 12$ m
  Sides 3 and 4 added together: $2 \times 5.5 = 11$ m
  $12 + 11 = 23$ m of facia
2 Number of pipe units needed: $5.4 \div 0.6 = 9$

PAGE 49

The individual dimensions (650 mm, 2800 mm, 650 mm) add up to an overall dimension of 4100 mm. The diagram incorrectly states the overall dimension as being 4000 mm.

Glossary

Aerated: exposed to the circulation of air

Aggregate: mineral material in the form of grains or particles, typically made of sand, stone, gravel, recycled concrete and crushed rock

Arris: the long, straight, sharp edges of a block or brick formed at the junction of two faces

British Standards: the UK authority that sets out a range of standardised quality requirements, procedures and terminology

Broken bond: the use of cut blocks (or bricks) to establish a good bonding pattern where full blocks (or bricks) will not fit

Building line: a boundary line set by the local authority beyond which the front of a building must not project

Carbon emissions: the release of carbon into the atmosphere (also referred to as greenhouse gas emissions) contributing to climate change

Carbon footprint: the amount of greenhouse gases – primarily carbon dioxide – released into the atmosphere by a particular human activity

Cladding: a covering or coating of one material over another to provide a skin or layer

Client: a person or company that receives a service in return for payment

Combustible: able to catch fire and burn easily

Compacted: firmly packed or pressed together

Compensation: something (usually money) awarded to someone in recognition of loss, suffering or injury

Compressive strength: the capacity of a material to resist pressure and squeezing forces without breaking

Conductor: a material through which electricity can flow freely

Conventions: an agreed set of standards, practices and methods for producing drawings

Courses: continuous rows or layers of bricks or blocks on top of one another

Cure: to set hard, often using heat or pressure

Damp-proof course: a continuous barrier built into masonry at specified locations to prevent moisture entering a structure

Datum: a reliable fixed point or height from which reference levels can be taken

Demolish: to deliberately destroy something (such as a building)

Dry silo mixer: a machine that contains all the dry materials to produce precise quantities of mixed mortar on demand

Duct: a tube or channel which allows the passage of a liquid or gas

Elements (of a building): the main parts of a building – the foundation, floors, walls and roof

Environment: the natural world as a whole, affected by human activity

Fabric (of a building): the structure of a building and the materials from which it is made

Face plane: the accurate alignment of all the blocks or bricks in the face of a wall to give a uniform flat appearance

Footing: the section of masonry from the concrete foundation to the ground-floor level; sometimes the whole foundation is referred to as 'footings'

Friction: the resistance that one surface or object encounters when moving over another

Frontage line: the front wall of a building

Gauge: in this context, the process of establishing measured uniform spacing between brick or block courses including the horizontal mortar joints

Hardcore: solid materials used to create a base for load-bearing concrete floors, paths or roadways

Hatchings: a standardised set of lines and symbols that allow easy identification of materials shown on a drawing

Hydrated: caused to heat and crumble by a chemical reaction when combined with water

Indents: holes or pockets accurately formed at each course or block of courses in the main wall as building proceeds

Induction: an occasion when someone is introduced to and informed about a new job or organisation

Insulated: covered in a material through which electricity cannot flow freely

Joists: parallel timber beams spanning the walls of a structure to support a floor or ceiling

Kiln: a type of large oven

Kinetic: relating to, caused by or producing movement

Legislation: a law or set of laws made by a government

Linear measurement: measuring the distance along a line between two points

Load-bearing: supporting the weight of a building or parts of a building

Method statements: documents giving information on safe and efficient sequences of work

Modular: designed with standardised units that can fit together in a variety of ways

Percentage: part of a quantity expressed in hundredths

Perimeter: in this context, the total linear measurement around the edge of a building

Perp: short for 'perpendicular'; the vertical mortar joint at right angles to the horizontal mortar bed joint

Personal protective equipment (PPE): all equipment (including clothing affording protection against the weather) which is intended to be worn or held by a person at work and which protects against one or more risks to a person's health or safety

Plan view: a view from directly above the subject – a 'bird's-eye' view

Plasticiser: an additive used to make mortar more workable and pliable

Prefabricated: factory-made units or components transported to site for easy assembly

Profiles: (in the context of setting out) timber boards and pegs assembled at the corners and other wall locations of a building to allow string lines to be positioned accurately

Proprietary: manufactured and sold under a brand name or trademark

Rafters: beams set to suit the angle of the roof pitch, forming part of the roof's internal framework

Ratio: the amount or proportion of one thing compared to another

Reinforcement: the process of strengthening something

Rendered: where a wall is coated with a sand/cement mix to provide a smooth, weatherproof finish

Return: expression used in bricklaying to describe the portion of blockwork (or brickwork) at right angles to the face of the wall

Reverse bond: for a wall with corners at each end, one end will have a block in line with the face of the wall and the other end will have a block at right angles (90°) to the face of the wall; for a wall with stopped ends, one end will have a full block and the other end will have a half block

Ridge: the highest horizontal line on a pitched roof where the sloping surfaces meet

Rising damp: when moisture from the ground travels up through the walls by capillary action

Scale: when accurate sizes of an object are reduced or enlarged by a stated amount

Schedule: a list of repeating components or features showing the building or site location where they are intended to be installed

Segregated: separated into groups or categories

Semi-permeable: material that allows only certain materials or substances (such as water vapour) to pass through it

Site induction: an occasion when someone is introduced to and informed about a new job or organisation

Specification (for construction): a detailed description of materials and requirements for a construction project

Standardised: conforming to a set standard

Stopped end: squared-off vertical finish to the end of a wall

Substructure: the section of a building extending below ground-floor level

Superstructure: the section of a building from the ground-floor level upwards

Sustainability: meeting our own needs without damaging the ability of future generations to meet their own needs

Tolerances: allowable variations between specified measurements and actual measurements

Tool-box talks: short meetings arranged at regular intervals at your work location on site to discuss safety issues; they give safety reminders and inform personnel about new hazards that may have recently arisen, for example, due to extreme weather

Topsoil: the upper, outermost layer of soil, usually the top 13–25 cm (5–10 inches)

Verbal communication: transmitting information by talking to others

Voltage: the amount of potential energy between two points in an electrical circuit, expressed as 'volts' or 'V'

Volume: the total space in three dimensions taken up by an object, material or substance

Well-graded sand: sand that has large, medium and small grains

Index